COLLECTIVE PHENOMENA IN CURRENT-CARRYING PLASMAS

COLLECTIVE PHENOMENA IN CURRENT-CARRYING PLASMAS

E. D. Volkov, N. F. Perepelkin,
V. A. Suprunenko, and E. A. Sukhomlin

*Khar'kov Institute of Physics and Technology,
Ukrainian SSR Academy of Sciences,
Khar'kov, USSR*

Translated from the Russian by
Dave Parsons

GORDON AND BREACH SCIENCE PUBLISHERS
NEW YORK LONDON PARIS MONTREUX TOKYO

Gordon and Breach Science Publishers

P.O. Box 786
Cooper Station
New York, New York 10276
United States of America

P.O. Box 197
London WC2 4DL
England

58, rue Lhomond
75005 Paris
France

P.O. Box 161
1820 Montreux-2
Switzerland

14-9 Okubo 3-chome,
Shinjuku-ku,
Tokyo 160
Japan

Originally published in Russian as
Коллективные явления в токонесущей плазме
by Izdatel'stvo Naukova dumka, Kiev, 1979

Library of Congress Cataloging in Publication Data

Kollektivnye ĭavlenĭia v tokonesushcheĭ plazme. English.
Collective phenomena in current-carrying plasmas.

Translation of: Kollektivnye ĭavlenĭia v tokonesushcheĭ plazme.
At head of title: Khar'kov Institute of Physics and Technology, Khar'kov, USSR. Ukrainian SSR Academy of Sciences.
1. Plasma turbulence. 2. Plasma (Ionized gases)—Conductivity. I. Volkov, Evgeniĭ Dmitrievich. II. Khar'kovskiĭ fiziko-tekhnicheskiĭ institut.
QC718.5.T8K6513 1985 530.4'4 84-21116

ISBN 2-88124-016-X. Printed in Great Britain by Bell and Bain Ltd., Glasgow.

To our teacher,
Academician
Evgeniĭ Konstantinovich Zavoĭskiĭ

CONTENTS

FOREWORD

It is now about a quarter of a century since the united international world of plasma physicists first heard exciting sounds from the Kurchatov Institute of Atomic Energy, Moscow, and the Khar'kov Physical Technical Institute (now the Khar'kov Institute of Physics and Technology) of the Ukrainian SSR Academy of Sciences announcing the achievement of reaching unprecedented temperatures in plasmas at densities relevant for fusion research. Relatively simple techniques had to be used. The concept of "turbulent heating" was born, proposed by the inspiring Academician Evengiĭ Konstantinovich Zavoĭskiĭ (1907–1976). Since then the name "turbulent heating" has been universally adopted. The important discoveries in these experiments were rightly officially rewarded in 1973 by the Committee on Discoveries and Invention of the USSR Council of Ministers.

The occurrence of collective processes had already been noted much earlier in the observation of superthermal radio emission from solar activity near the fundamental and first harmonic of the plasma frequency (1940–1950). Later, in experiments where a beam was injected, comparable superthermal radioemission spectra could be provoked near the characteristic plasma frequencies and their higher harmonics.

Following the theoretical predictions of Budker and Buneman, probably the first purposeful excitation of a current-driven turbulent plasma state was performed in a straight discharge called "GROM-2" (1963) and communicated in a paper by some of the present authors. It is high time that the English translation of this Russian monograph on the subject of collective phenomena in current-carrying plasmas written by members of the famous Khar'kov group be completed. I should also like to regard the appearance of the English version as a memorial to our friend and inspiring colleague in the study of plasma turbulence, the late Prof. Viktor Afanas'evich Suprunenko (1932–1983), toroidal confinement section head at the Khar'kov Institute of Physics and Technology.

The authors deserve our thanks for this comprehensive study of experimental results of plasma turbulence studies carried out in diverse devices all over the world. A system of classifying various turbulent states derived from their own and international research is presented; in its essentials, it is purified from the complicating boundary conditions. A classification is shown to be possible in terms of the particular ratios E/E_{Dr} (where E_{Dr} is the well-known runaway electric field strength), $(\omega_{ce}/\omega_{pe})^2$, $V_{drift}/v_{thermal}$, and T_e/T_i pertaining to a plasma state. For various combinations of these ratios, the real and imaginary parts of the collective conductivity, electron and ion heating rates, and superthermal spectra have been explored in wide density regions. The results are shown in numerous graphs and tables, giving a fair exposition of the state of knowledge in this interesting area of plasma physics research.

The rationalization of the phenomenology of plasma turbulence was only possible through the support and interaction of the formidable theoretical work through the years. Though a very satisfactory agreement between predictions of macroscopic variables and theory was often obtained, giving confidence in the conceptions used, the authors have justifiably given very thorough attention to the study of the microinstabilities proper. With the use of modern diagnostic methods, the various wave processes and their interactions, which have been postulated as necessary ingredients in the theoretical concepts, have been identified in many cases. The need for further exploration is, however, clearly brought out.

The potential of turbulent heating to produce "thermonuclear grade" plasmas in large devices has also been discussed and, in principle, supported. It is, however, realized that certain favorable conditions, e.g., sufficient radial energy transport, have to be present in order to make applications to large systems possible. Positive signs have been found in various experiments that indeed indicate the presence of that transverse transport of wave energy. Inspired by the relative ease of obtaining extremely high temperatures and high kinetic pressures in smaller devices, it remains a desire of researchers in this field to test one of the turbulent heating schemes in one of the large present-day machines.

There is no doubt that even those investigators who would rather avoid a confrontation with turbulence in their plasma studies are often confronted with unwanted collective phenomena. This statement can easily be confirmed by reading the subjects listed in any arbitrary plasma physics journal.

Fundamental research on turbulence in plasmas continues. Many of the different forms of turbulence shown in this book are isolated in the purest possible form in order to be investigated systematically and in great depth. These activities will doubtless give the authors a well-deserved sense of satisfaction.

Hans de Kluiver,
FOM-Instituut voor Plasmafysica,
Rijnhuizen, Nieuwegein, Nederland

PREFACE

Many years of research on gas discharges have demonstrated the importance of collective interactions in the processes which occur in nonequilibrium plasmas. These collective interactions are manifested in a wide variety of instabilities, which may be driven by several factors: a beam of charged particles in the plasma, a current in the plasma, or a spatial inhomogeneity of the plasma density and temperature in the discharge.

This monograph analyzes collective phenomena, anomalous resistance, and turbulent heating in a current-carrying plasma. The primary goals are to put in systematic form and critically review the experimental work which has been done in various labortories around the world and to identify the phenomenological laws which determine the course of the events of interest in specific experimental systems.

Experimental work on anomalous resistance and turbulent heating has grown considerably since the appearance (in 1967) of the last review on the subject, by Zavoĭskiĭ and Rudakov.[91] Turbulent ion heating has been studied; a fast tail has been found on the ion distribution function; the dependence of the turbulent level and the stabilization of the drift velocity on the magnetic field has been studied; the turbulence spectrum has been studied in detail; the mechanism for the superthermal rf emission from plasmas has been identified, making it possible to resolve the contradiction between the approximately steady nature of the anomalous plasma resistance and the sporadic nature of the superthermal rf emission; it has been shown that a turbulently heated plasma can be confined; and the loss mechanisms during turbulent heating have been studied.

On 5 November 1973, the Committee on Discoveries and Inventions of the USSR Council of Ministers awarded Certificate No. 112 for the discovery of "Turbulent heating and the anomalous resistance of plasmas,"[22] to a team of researchers led by Academician Zavoĭskiĭ from the Kurchatov Institute of Atomic Energy, Moscow, and the Khar'kov Physicotechnical Insitute of the Ukranian Academy of Sciences. The creative early work in the field by Zavoĭskiĭ, the organizing effect of his work, and his continual generation of new ideas and interest in them determined the subsequent development of the field and drew a widespread response from the world scientific community. The term "turbulent heating," proposed by Zavoĭskiĭ, has been adopted universally.

An effort is made in this monograph to generalize the experimental results on collective interactions in current-carrying plasmas which have been obtained over the past 20 years in laboratories throughout the world, to show how some results which appear contradictory at first glance actually fit in the overall picture, and to suggest some major problems for future study. It is the hope of the authors that a comprehensive analysis of the work which has already been done will lead to a better understanding of the anomalous resistance and turbulent heating of plasmas and will be of assistance to future development of the theory.

We wish to thank all of our colleagues for furnishing experimental data, and we thank O. A. Bezruk and V. G. Kaz'yanenko for assistance in putting the manuscript together.

Please note that the authors' names on the title page are ordered in accordance with the Russian alphabet.

E. A. Volkov
N. F. Perepelkin
V. A. Suprunenko
E. A. Sukhomlin

BASIC NOTATION

H: magnetic field, Oe

V_0: potential difference between the electrodes of a straight discharge, V

E_0: vacuum electric field in the discharge, V/cm

E: electric field in the plasma, V/cm

$E_{Dr} = 0.67 \cdot 10^{-12} \frac{n}{T}$: Dreicer critical electric field for hydrogen, V/cm

I: current, A

j: current density in the plasma, A/cm^2

$T_e = \frac{m_e v_{Te}^2}{2}$: electron temperature of the plasma, eV

$T_i = \frac{M_i v_{Ti}^2}{2}$: ion temperature, eV

$T = T_e + T_i$: total plasma temperature, eV

$v_{Te} = 5.93 \cdot 10^7 \sqrt{T_e}$: electron thermal velocity, cm/s

$v_{Ti} = 1.38 \cdot 10^6 \sqrt{T_i}$: ion thermal velocity, cm/s

v_b: velocity of electron beam injected into the plasma, cm/s

$v_S = 0.98 \cdot 10^6 \sqrt{T_e}$: ion acoustic velocity, cm/s

v_φ: phase velocity of wave, cm/s

$v_A = \frac{H}{\sqrt{4\pi M_i n}} = 0.22 \cdot 10^{12} \frac{H}{\sqrt{nA}}$: Alfven velocity, cm/s

$c = 2.9949 \cdot 10^9$: speed of light, cm/s

$u = \frac{I}{S_{\text{eff}} ne}$: electron current drift velocity, cm/s

$\alpha = u/v_S$: measure of the stabilization of the current drift velocity

R: major radius of torus, cm

a: minor radius of plasma, cm

l: distance between electrodes of straight discharge, cm

λ_0: wavelength of the microwave probe, cm

λ: wavelength of a wave excited in the plasma, cm

$\mathbf{k}$: wave vector, cm^{-1}

$\mathbf{k}_s$: wave vector of ion acoustic wave, cm^{-1}

k_0: scale dimension of turbulence, cm^{-1}

$\mathbf{k}_l$: wave vector of a plasma wave (Langmuir wave), cm^{-1}

$\rho_e = \frac{v_{Te}}{\omega_{ce}} = 3.37 \frac{\sqrt{T_e}}{H}$: electron Larmor radius, cm

$\rho_{ci} = \frac{v_{Ti}}{\omega_{ci}} = 144 \frac{\sqrt{T_i}}{H}$: ion Larmor radius, cm

$\rho_D = \sqrt{\frac{T_e}{8\pi n e^2}} = 530 \sqrt{\frac{T_e}{n}}$: Debye length, cm ($\rho_D \equiv \lambda_D$)

$m_e = 9.1085 \cdot 10^{-28}$: electron mass, g

$M_H = 1.6734 \cdot 10^{-24}$: mass of hydrogen atom, g

M_i: ion mass

$M_H/m_e = 1837$

$A = \frac{M_i}{M_H}$: ratio of the ion mass to the mass of the hydrogen atom

$e = 1.6 \cdot 10^{-19}$: electron charge, C

Z_{eff}: effective charge of the plasma ions, in units of the electron charge

$\omega_{pe} = \left(\frac{4\pi n e^2}{m_e}\right)^{1/2} = 5.64 \cdot 10^4 \sqrt{n}$: electron plasma frequency, s^{-1}

$\omega_{pi} = \left(\frac{4\pi ne^2}{M_i}\right)^{1/2} = 1.32\cdot10^3\sqrt{n}$: ion plasma frequency for a hydrogen plasma, s^{-1}

$\omega_h = \sqrt{\omega^2_{pe} + \omega^2_{ce}}$: upper hybrid frequency, s^{-1}

ω_0: frequency of electromagnetic probe beam, s^{-1}

$\omega_{ce} = \frac{eH}{m_e c} = 1.76\cdot10^7 H$: electron cyclotron frequency, s^{-1}

$\omega_{ci} = \frac{eH}{M_i c} = 9.58\cdot10^3 \frac{ZH}{A}$: ion cyclotron frequency, s^{-1}

$$\omega^* = \frac{cT_e}{eH}\, k \left(\frac{1}{n}\ \frac{\partial n}{\partial r}\right) = 10^8\ \frac{T_e}{H}\ \frac{k_\perp}{a} \left(\frac{a}{n}\ \frac{\partial n}{\partial r}\right):$$

characteristic frequency of the drift instability, s^{-1}

Γ: growth rate of an instability in the plasma, s^{-1}

γ: intensity of hard x-ray emission

$\nu_{ee} \approx \nu_{ei} = 2.9\cdot10^{-5} n \frac{Z}{T_e^{3/2}} (\Lambda/10)$: rate of binary collisions of electrons with ions and electrons, s^{-1}

$\nu_{ii} = 4.7\cdot10^{-7} \frac{Z^3 n}{T_i^{3/2}}\ \frac{(\Lambda/10)}{A^{1/2}}$: rate of binary collisions of ions with ions, s^{-1}

$\Lambda = 25.3 - 1.15\log n + 2.3\log T_e$ (for $T_e > 50$ eV): Coulomb logarithm

R_{eff}: resistance of plasma column

$$\tau_{eq} = \tau_{ei}\left(\frac{M_i}{3m_e}\right) = 2.11\cdot 10^7 \frac{T_e^{3/2} A}{(\Lambda/10) Zn}$$: scale time for energy transfer from electrons to ions

τ_0: scale time for turbulent heating of the plasma in a straight discharge

τ_E: energy lifetime of the plasma

τ_E^*: lifetime of the fast particles in a closed magnetic confinement system

τ_n: particle lifetime in a magnetic confinement system

D: particle diffusion coefficient

$$\chi_{\parallel}^e = 1.58\,\frac{nv_{Te}^2}{\nu_{ee}}$$: electron thermal conductivity of the plasma along the magnetic field

$$\chi_{\perp}^e = 2.33 n\rho_{ce}^2 \nu_{ee}$$: electron thermal conductivity of the plasma across the magnetic field

n: plasma density

n_0: plasma density before the turbulent-heating current begins to flow

$$\sigma_{\perp} = \frac{\omega_{0e}^2}{4\pi\nu_{ei}} = \frac{10}{\left(\frac{\Lambda}{10}\right)}\,\frac{T_e^{3/2}}{Z}$$: Coulomb electrical conductivity of a plasma across the magnetic field, mho/cm

$\sigma_{\parallel} \mp 1.96\sigma_{\perp}$: Coulomb conductivity of the plasma along the magnetic field, for hydorgen, mho/cm

σ_{turb}: turbulent electrical conductivity of the plasma, mho/cm

W_l: energy density of the plasma waves (Langmuir waves) in the plasma

W_s: energy density of the ion acoustic waves in the plasma

$W_{\Omega,\vartheta}$: energy density of plasma whistlers

Ω: frequency of plasma whistler

ϑ: angle between the wave vector of a wave and the magnetic field

$f(v)$: velocity distribution function of the plasma particles

Note: Where given, the numerical factors correspond to a hydrogen plasma.

CHAPTER I
ELECTRIC CURRENT FLOW IN PLASMAS

1 *INTRODUCTION*

The long-range nature of the Coulomb attraction and repulsion, the large number of degrees of freedom, and the effective interactions of the charged particles with electric and magnetic fields give rise to the collective behavior of a plasma and distinguish it as a fourth state of matter. Although the plasma state is in fact the most common one in nature, it is also the one which has received the least study, at least until recently.

Experimental research on the physics of hot plasmas began to develop rapidly when the problem of controlled thermonuclear fusion was taken up. The first experiments on this problem showed that plasmas do not conform to the simple laws predicted by the classical theories based on binary collisions. Plasmas proved to be subject to many instabilities, which prevented effective heating and prolonged confinement.

The revolutionary idea of turning the microinstabilities of a current-carrying plasma to advantage, and using them to achieve effective heating of the plasma, was advanced by Sagdeev in a theory which he developed for shock waves in a low-density plasma. Turbulent current heating was first achieved experimentally at laboratories in Moscow and Khar'kov in the early 1960s and has proved to be an effective method for producing hot plasmas in magnetic confinement systems.[17,19,159,160] Underlying this heating method is the anomalously high resistance exhibited by a current-carrying plasma as the current interacts with waves which it itself excites in the plasma. No fundamental limits are imposed on the temperature or the density for this heating method, and the plasma can be heated quickly, before catastrophic instabilities can occur in the discharge. A necessary condition for turbulent heating is that the current density must be kept above a critical value.

The widespread development of experimental and theoretical research which ensued was rewarded with a deeper understanding of the physics of current-carrying plasmas, and it explained many of the phenomena observed in nature and in the laboratory. In particular, the disruption of the electron acceleration in a plasma betatron turned out to result from the Budker–Buneman instability. The same mechanism is probably responsible for certain types of radio bursts from the sun.

The extensive program of research on turbulent current heating of plasmas in various laboratories around the world had a primary goal of making direct practical use of the results of fundamental research on the turbulent state of plasmas in order to achieve approximately steady-state turbulent current heating of plasmas in large fusion devices.

Resolving whether this heating method will be successful in fusion devices primarily requires studying its effectiveness in strong magnetic fields, in which the plasma is fundamentally magnetized, and the condition $\omega_{ce}^2/\omega_{pe}^2 \gg 1$ holds. It is simple to show that the equilibrium condition for a plasma in a magnetic confinement system, $8\pi nT/H^2 \leq \beta_{cr}$, is identical to the condition $T \leq (mc^2/2)\,(\omega_{ce}/\omega_{pe})^2\,\beta_{cr}$. For most closed magnetic confinement systems the relative equilibrium pressure is low, $\beta_{cr} \leq 10^{-2}$. For fusion temperatures on the order of 10^8 K, the ratio of the electron gyrofrequency to the electron plasma frequency is thus greater than unity; i.e., the plasma is magnetized.

Experiments have shown that the current drift velocity is stabilized in this case, as in a nonmagnetized plasma, but at a level about an order of magnitude higher. Successful turbulent heating in the magnetized plasma of a fusion device will therefore require a much higher current density.

Plasma confinement in a closed magnetic confinement system with approximately steady turbulent heating was first achieved in the URAGAN stellarator. The same experiments revealed an anomalous electron thermal conductivity, which could be described well by the pseudoclassical expression which Artsimovich had proposed in an effort to explain experiments on the ohmic heating of plasmas in tokamaks. There is a point to be made here which is extremely important in connection with the nature of the collective interactions: The effective collision rate in this pseudoclassical expression for the thermal conductivity is determined from measurements of the electrical conductivity of the plasma, while this circumstance would generally be obvious only for Coulomb collisions. Another interesting point is that in a gradual transition from a Coulomb conductivity to a turbulent conductivity the numerical factor in the pseudoclassical expression remains constant. This constancy means that the turbulent collision rate, which is of a collective nature, plays a role here which is similar to that played by binary collisions.

The use of turbulent heating to produce a dense plasma in the GROM-4 straight mirror system led to record high plasma properties for such systems: gas pressures up to 10^{19} eV/cm^3 at plasma densities up to 10^{15} cm^{-3}. After the turbulent-heating current was turned off in these experiments, the plasma was confined for a time which was determined by the properties of the confinement system itself and which agreed well with theoretical predictions. The turbulent heating produced a plasma with a hot core; within this core, the cyclotron instability, which limits the lifetime in ordinary mirrors, was suppressed. Experiments with deuterium plasmas in the same device revealed a neutron flux corresponding to a plasma temperature of 10^7 K. This work not only demonstrated that turbulent heating could be used to produce plasmas in mirrors but also rekindled the hope that confinement systems of this type could be developed into fusion reactors.

Study of the energy loss caused by electrons in straight systems revealed that this loss is determined by the classical thermal conductivity along the magnetic field when the turbulent collision rate is considered. Consequently, while in open systems the elevated turbulent collision rate reduced the energy loss due to electrons and thus substantially increased the attainable pressures, in closed magnetic systems (with the pseudoclassical anomalous loss) the turbulent heating simply increased the rate at which the maximum pressure was attained.

It is worthy of note that at high plasma densities and temperatures, at which the pseudoclassical loss law does not hold, steady-state turbulent heating may prove extremely valuable in closed magnetic confinement systems.

The spectrum and level of the turbulence in a current-carrying plasma are crucial to a determination of the physical nature of the anomalous resistance and turbulent heating of plasmas. Recent experiments which were carried out to analyze the local trapping and buildup of weakly damped rf electromagnetic noise in bounded regions in an inhomogeneous plasma have resolved the apparent contradiction between the sporadic nature of the superthermal rf emission from a plasma and the approximately steady-state nature of the anomalous resistance.

The local buildup of electromagnetic energy in a plasma is of fundamental importance to the problems of controlled thermonuclear fusion and magnetic confinement of plasmas, and it furthermore explains many effects, e.g., the explosive release of electromagnetic energy and the acceleration of charged particles in the plasma of the solar corona.

Turbulent current heating of plasmas has been used widely throughout the world to produce hot plasmas in large fusion devices, e.g., the Texas Turbulent Tokamak (USA), the TORSO stellarator (UK), the Nagoya Turbulent Mirror (Japan), the 2X2 mirror at Livermore (USA), the turbulent-heating tokamak at Frascati (Italy), and the OGRA-2 turbulent-heating mirror at the Kurchatov Institute in Moscow.

2 CLASSICAL PICTURE OF A CURRENT IN A PLASMA

Even in the early days of research on plasmas as quasi-neutral media consisting of charged particles of different signs, much effort was devoted to learning about the flow of electric currents in plasmas and about the nature of the resistance of plasmas. This research was based on two assumptions: that the interaction between the particles in the plasma is a Coulomb interaction and that the electric field is weak. The field was assumed weak in the sense that the energy acquired over a mean free path is much smaller than the thermal energy.

Let us examine the classical approach to the problem.[8,7,146,154] If an electric field E is produced in a plasma, it will put the charged particles in motion and thus give rise to an electric current. In general, the current carriers are electrons and ions, but it is much more difficult to move the plasma ions (in the direction opposite that of the electrons) because of their much higher mass. If the electrons and ions did not collide with each other, the electric field would continuously increase their velocities, and the plasma would eventually be transformed into two oppositely directed streams of fast particles. In reality, however, collisions occur and cause a different situation. Each particle acquires a directed velocity only during a limited time interval. During a collision, the additional velocity which has been acquired is lost, and after the collision the field begins to accelerate the particle again.

The Coulomb interactions are extremely unusual. The attractive and repulsive Coulomb forces are long-range forces and remain appreciable even when the charged particles are moved far apart (despite the fact that the force falls off rapidly, in proportion to $\frac{1}{r^2}$). As a result, most of the scattering is caused by a large number of deflections through small angles during collisions of the particles. These small deflections add up in accordance with statistical laws and gradually cause important changes in the momentum, energy, and coordinate of a charged particle. Deflections through large angles are improbable.

Although the basic change in the velocity of the charged particles is by no means analogous to the changes which occur in short-range collisions of neutral particles in a gas, it is still customary to introduce the concepts of mean free paths and mean free times between two successive collisions in a plasma. In particular, the mean time interval for electron-ion collisions, τ_{ei}, is the time required for the velocity direction to change by 90^0. This effective mean time is used for quantitative calculations.

When a current is flowing in a plasma, the additional velocity acquired by an electron and an ion over one mean free time is very small in comparison with the thermal velocity if the electric fields are weak. The reason is that in the steady state the maximum directed electron velocity is determined by the balance between the energy acquired in the electric field and the energy transferred through Coulomb collisions.

Since the velocity of the directed motion of the charged particles (the drift velocity) in a plasma is small in comparison with the steady-state velocity of the random (thermal) motion, the overall motion of the particles is changed only very slightly when an electric field is applied. A small velocity component parallel to the electric field is superimposed on the fast random motion of the particles at the thermal velocities. One may thus say that there is a comparatively slow drift of the particles along the field direction (a "shifted Maxwellian distribution"), rather than speak of the formation of a stream in which the velocities of all the particles are in the same direction.

In examining current flow in a plasma we must emphasize that the contribution of the (heavy) ions is very small, since they acquire very small directed velocities from the field. Most of the current in a plasma is carried by electrons. All the calculations, therefore, consider only the directed motion of the plasma electrons.

Under steady-state conditions, in a constant electric field E, the force exerted on the electrons must be balanced by the frictional force caused by collisions of the electrons with ions. This force is equal to the momentum lost by an electron per unit time through these collisions. In 1 s, an electron of mass m_e and charge e undergoes $\frac{1}{\tau_{ei}}$ collisions, during each of which it loses a directed velocity u and transfers a momentum $m_e u$ to ions. The frictional force is $\frac{m_e u}{\tau_{ei}}$, and the force equilibrium is expressed by the condition

$$eE=\frac{m_e u}{\tau_{ei}}\ldots \tag{I.1}$$

To determine the electrical conductivity of the plasma, we write an expression for the current density, j, in a plasma with an electron number density n: $j = neu$. Substituting in the value of u from (I.1), we find

$$j = \frac{ne^2}{m_e} \tau_{ei} E,$$

or

$$\sigma = \frac{ne^2 \tau_{ei}}{m_e}.$$

This relationship between j and E means that the current in the plasma obeys Ohm's law.

The electrical conductivity of a plasma is frequently expressed in terms of the electron plasma frequency ω_{pe} and the collision rate $\nu_{ei} = 1/\tau_{ei}$:

$$\sigma = \omega_{pe}^2/(4\pi\nu_{ei}) \quad . \; . \; . \qquad (I.2)$$

After determining the functional dependence of ν_{ei} on the electron density n and the electron temperature T_e of the plasma, we find the final expression for the electrical conductivity of the plasma.

This simplified approach yields an approximate value for the conductivity. A comprehensive derivation incorporating the electron velocity distribution and electron–electron collisions was carried out by Spitzer,[155] who showed that the conductivity σ of a fully ionized plasma is proportional to $T_e^{3/2}$, if a magnetic field is present the ratio of the longitudinal ($\sigma_{\parallel}$) and transverse ($\sigma_{\perp}$) components with respect to the magnetic field is 1.96 for $Z = 1$.

Braginskii[41] has carried out the most accurate calculation of the plasma conductivity, determining the numerical coefficients:

$$\sigma_{\parallel} = \frac{1{,}76\cdot 10^{13}}{(\Lambda/10)Z} T_e^{3/2} (\mathrm{s}^{-1}\cdot \mathrm{eV}^{-3/2}) = 19{,}6 \frac{T_e^{3/2}}{(\Lambda/10)Z} (\mathrm{mho/cm}),$$

where Z is the ion charge. It can be seen that the conductivity of a fully ionized plasma is essentially independent of its electron density. The physical explanation for this result is that an increase in the electron density is cancelled precisely by an equal increase in the ion density and simultaneously leads to an increase in the number of collisions, i.e., to a decrease in the mean free path. The conductivity of a fully ionized plasma decreases with increasing average ion charge and increases in proportion to the electron temperature raised to a power of 3/2. At high temperatures the conductivity can become very high. At T_e = 1 keV, for example, the conductivity of a hydrogen plasma is nearly equal to that of copper at room temperature.

This finite conductivity of plasmas is exploited in the ohmic method of heating plasmas, by means of electric currents. Here the energy of the electric field is ultimately converted by binary collisions into the thermal energy of (primarily) the plasma electrons. Collisions of electrons with ions serve to randomize the directed velocity, while electron—electron collisions cause the temperature of the electron gas to rapidly become uniform. Energy exchange between electrons and ions occurs slowly. The efficiency of ohmic heating, however, falls off sharply with increasing plasma temperature. This effect reflects the energy dependence of the Coulomb cross sections of the charged particles; the cross section for the Coulomb interaction of electrons with ions falls off with increasing electron velocity in proportion to $1/v^4$.

The nature of these interactions means that acceleration occurs in plasmas even in weak electric fields. The conditions for binary interactions of the electrons in different parts of the distribution function are quite different. For fast electrons (from the high-energy tail of the Maxwellian distribution) the mean free path (which increases in proportion to v^3) is much longer than that for slow electrons ($v \geq v_{Te}$). The dynamic frictional force due to electron–ion collisions falls off in proportion to $1/v^2$, so that these fast electrons may go into continuous acceleration if the force exerted on them by the electric field exceeds the dynamic frictional force. Even in weak electric fields there is always a group of electrons in the Maxwellian distribution which become "runaway" electrons. In this case the only electrons which are caught up in the acceleration are those from the extreme high-energy end of the distribution function. Collisions smooth out the distortions of the original electron distribution function, and the electrons lost from the part of the distribution function involved in the acceleration are continuously replaced. There is accordingly a gradual transition of the bulk of the plasma electron distribution into the region where the electrons become runaways. In principle, after a sufficiently long time in an unbounded plasma, the bulk of the electrons could become runaways, if no non-Coulomb mechanism acted to limit the process.

We have been discussing the case of weak electric fields, in which the directed energy acquired by electrons in the time between collisions is less than the thermal energy of the plasma. In the case of strong electric fields the opposite is true: The energy acquired by an electron from the electric field over a mean free path is greater than the average energy of the random motion of the plasma electrons. In this case most of the plasma electrons become runaways. A critical electric field is specified by the condition that at this field value the entire current in the plasma is carried by electrons with a directed velocity equal to the thermal velocity:

$$j_{cr} = nev_{Te} = \sigma E_{Dr},$$

from which we find

$$E_{Dr} = nev_{Te}/\sigma .$$

Expressing v_{Te} and σ in terms of T_e, we find $E_{cr} = 0.67 \cdot 10^{-12}\, n/T_e$ for $Z = 1$ and $\Lambda = 14$ (for a rather hot hydrogen plasma).

The most comprehensive calculations on electron runaway have been carried out by Dreicer,[81] who studied the time evolution of the electron velocity distribution in an unbounded plasma in an electric field, calculated the electron diffusion time in velocity space in the runaway region for a given E, and determined the critical field. Dreicer takes the critical field to be the lower boundary of the energy range of the electrons which have gone into continuous acceleration. Below this boundary, the only electrons which become involved in the runaway are those in the Maxwellian tail, and the plasma contains, along with a current obeying Ohm's law, a current of accelerated electrons. As the critical field is approached the fraction of the electrons which are accelerated increases. Above E_{Dr} the current is carried exclusively by the accelerated electrons.

Harrison's studies[228,229] of runaway electrons were among the earliest studies of the topic, which included calculations, experimental detection of runaway electrons, and study of these electrons. Harrison obtained dense streams of runaway electrons in a straight system and carried out some rough calculations which pointed to a fundamental distinction between the runaway effects in bounded and unbounded plasmas. In toroidal systems, according to those calculations, the runaway-electron current can increase without bound (in proportion to the square of the time over which the electric field is applied), while in bounded straight systems the current is proportional to the length of the system, and it becomes independent of the time after saturation sets in.

The classical approach to current flow in a plasma thus gives rise to the concepts of a Spitzer conductivity, ohmic heating, and electron runaway. Most of the results reported here refer to the case of weak electric fields.

CHAPTER II
OBSERVATION OF THE ANOMALOUS ELECTRICAL CONDUCTIVITY OF PLASMAS

1 ANOMALOUS CONDUCTIVITY OF PLASMAS

The electrical conductivity of a plasma, one of the most important transport coefficients, reflects the fundamental physics of current flow in a plasma and the conversion of the energy of the directed motion of the electrons into random energy.

We have seen that in the classical case (of weak electric fields) the conductivity is determined by Coulomb collisions and is proportional to $T_e^{3/2}$. This conductivity is the basis for the ohmic heating of plasmas (with electric currents). As the electric field is increased, the current in the plasma rises, and the ohmic heating raises the temperature of the plasma and thus its conductivity.

As the electric field is strengthened, the conductivity would therefore be expected to increase. In 1959, however, it was discovered experimentally[40] that as the external electric field is strengthened the plasma conductivity does not increase; on the contrary, it decreases (Fig. 1), in explicit contradiction of the classical interpretation. (This discovery was made in experiments in the GROM-1 device, a high-current straight discharge with a skin effect in hydrogen, stabilized by a longitudinal magnetic field.) This unexplained effect was subsequently termed an "anomaly" of the plasma conductivity, and the conductivity itself was termed "anomalous." A more careful study of this effect showed that, among all possible causes, the leading role was played by the electric field in the plasma. It was shown that when this field is very weak (in comparison with the Dreicer field) the plasma conductivity is described approximately by the classical expression. In a strong field, processes observed in the plasma reflect the fundamental properties of a plasma as a quasineutral ensemble of charged particles: collective interactions of wave fields with particles. These interactions determine the magnitude and behavior of the plasma conductivity.

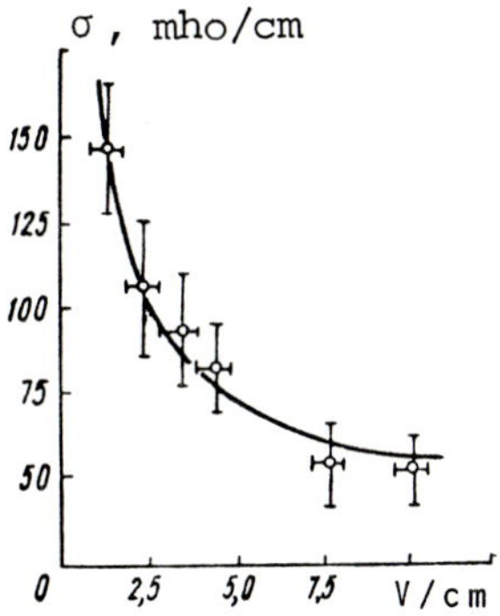

FIGURE 1 Electrical conductivity of the plasma vs. the electric field in a high-current gas discharge with a skin effect (the GROM-1 device[40]).

The unambiguous relationship between the onset of the anomalous conductivity and the excitation of intense microwave oscillations in the plasma was first established experimentally in 1961, in the GROM-3 device.[159] This relationship was subsequently found in other devices, in experiments on collective interactions in plasmas in strong electric fields. Figure 2 (from Ref. 161) demonstrates the effect; Fig. 3 shows some corresponding

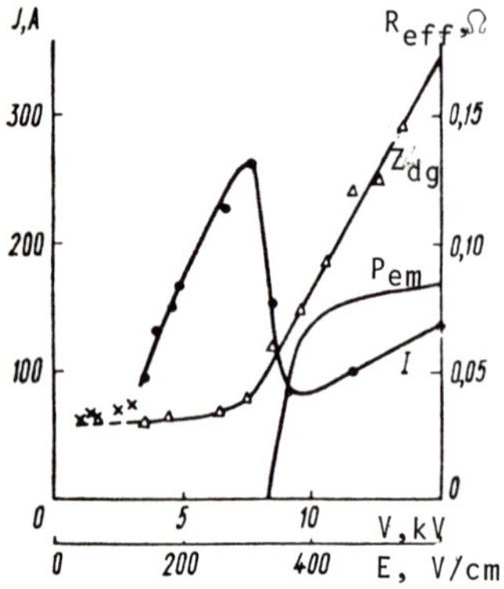

FIGURE 2 Resistance of the discharge gap, current density at the discharge axis, and microwave emission from the plasma vs. the electric field in the plasma (the GROM-3 device[159,161]).

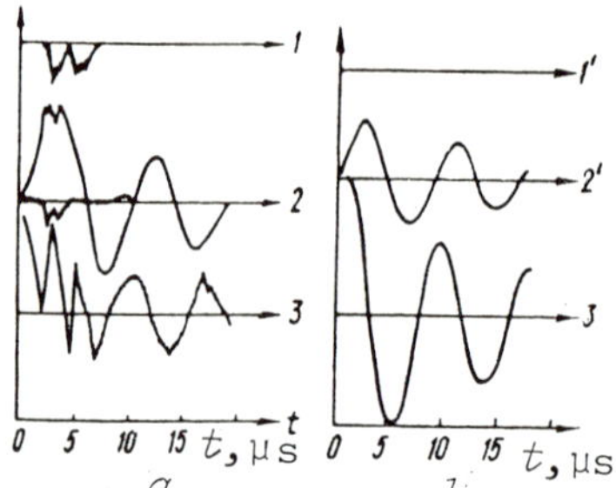

FIGURE 3 Oscilloscope traces of the microwave emission from the plasma (1, 1') of the total discharge current (traces 2 and 2', directed upward), the current of accelerated electrons (trace 2, directed downward), and the voltage across the discharge gap (3, 3') according to experiments in the GROM-3. *a*) Excitation of collective interactions; *b*) without collective interactions (depending on the level of the electric field with respect to the Dreicer field[160]).

oscilloscope traces. It can be seen from Fig. 2 that there is a threshold electric field, below which no microwave emission from the plasma is observed. Above this threshold field, intense microwaves are excited in the plasma, and an external receiver can detect electromagnetic emission. Simultaneously, there is a sharp increase in the resistance of the plasma, and the current corresponding to the directed electron drift is cut off. When the collective interactions come into play, the directed-motion energy which the electrons acquire in the external electric field is thus expended on exciting the small-scale, high-frequency oscillations which give rise to the anomalously high plasma resistance. It has been found experimentally that the pulsed emission power is high (reaching hundreds of kilowatts), and the radiated waves span a broad wavelength interval — from the submillimeter range to the meter range.

The mechanism for the transfer of directed-motion energy of the electrons to waves is thus very efficient, and the waves contain a substantial fraction of the energy provided by the external source.

This experiment established the most general properties of the anomalous conductivity as a physical phenomenon: its appearance in a sufficiently strong electric field, its correlation with the excitation of rf waves in the plasma, and the stabilization of the directed-drift velocity of the electrons at some level characteristic of the given conditions. This level is determined by which particular current-driven instability occurs in the system (the ion acoustic instability, the Buneman instability, the current—drift instability, etc.), and it gives rise to an effective dynamic-friction force of a collective nature.

Since the stabilization of the directed-drift velocity of the electrons results from the occurrence of some instability or other, we are led to the question of a critical electric field in the plasma and a critical current density. As the electrons are accelerated in the external electric field, they acquire a velocity near some characteristic velocity for the particular type of instability involved and excite waves in the plasma.

In straight discharges, where the electrons are accelerated a single time between the electrodes, the electric field would have to be very strong in order to raise the electron velocity to a level near the critical velocity. For this reason, in straight discharges the collective interactions are seen most clearly only in strong electric fields, above the Dreicer field. A sharp threshold is seen in the onset of the effect under these conditons.[288]

In toroidal systems, in contrast, collective interactions may also be manifested in weak electric fields, well below the Dreicer field.[45] The reason for the difference is that in toroidal systems the electrons can acquire a velocity near the critical velocity as they repeatedly orbit the torus. The process in a toroidal system may thus not have as clearly defined a threshold as in a straight system. In strong electric fields in a torus, the collective interactions are naturally more intense.

The reason why there is a critical current density for the onset of the current-driven instabilities is that, for a given plasma density, it is necessary to stabilize the directed electron velocity at a corresponding level. This condition means that the power supply must have an adequate power.

For a particular current-driven instability to occur in an experiment, two conditions must be satisfied simultaneously: one on the electric field ($E > E_{cr}$) and one on the current density in the plasma ($j > j_{cr}$). These conditions can be met simultaneously in high-current gas discharges.

High currents are produced in installations having energy banks which are operated in pulses. Pulsed operation presents some well-known difficulties in attempts to measure the plasma conductivity. In some experiments, therefore, where the conductivity anomaly is only weakly expressed, the investigators must deal with the problem of measuring this anomaly in the face of the errors of the measurement methods available. As an example we consider the complexity of determining the plasma conductivity in a magnetic-field-stabilized high-current straight discharge (the GROM-3 device[160]), in which the ratio of the classical conductivity to the anomalous conductivity amounts to several orders of magnitude.

The conductivity is found by measuring the pulsed current and the active component of the voltage. Measurements of the voltage (in contrast with the measurements of the current) present definite difficulties. In general, the total voltage across the pulsed discharge is given by

$V = IR_{act} + L\frac{dI}{dt} + I\frac{dL}{dt}$, where R_{act} is the resistance of the plasma, I is the current in the plasma, and L is the inductance of the discharge. All the difficulties arise in attempts to determine the inductive components. It is necessary to evaluate the inductance of the discharge and its time evolution. For this purpose a streak camera was used in the GROM-3 experiments to measure the diameter of the plasma column and its time dependence. Furthermore, since the emitting plasma volume may not always correspond to the current channel, the time evolution of the radial profile of the current density was determined by (magnetic) probe measurements. Figure 4 is a streak photograph showing the changes in the diameter of the plasma column. All the measurements were carried out in the time interval in which the minimum diameter of the plasma column did not change, i.e., in which the plasma was stable. The magnetic probe measurements showed that the current channel coincides with the emitting volume and that the radial profile of the current density does not change substantially over the time interval studied.

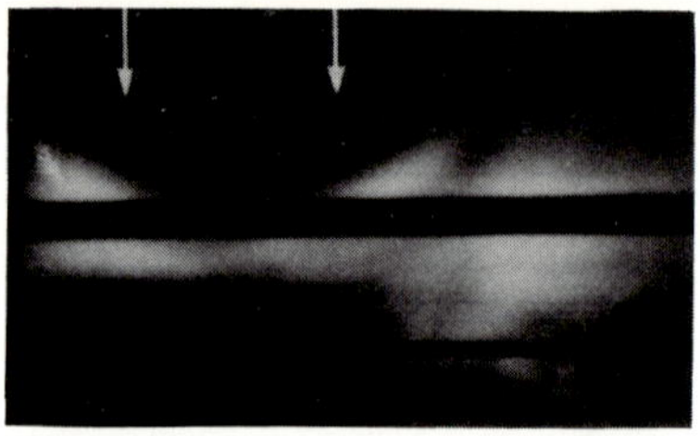

FIGURE 4 Streak photograph of the glow diameter of the plasma column in the GROM-3. The arrows show the time interval (on the order of 3 μs long) during which the anomalous plasma resistance, the microwave emission, and the heating are observed.

In this experiment the inductive components of the voltage turned out to be small in comparison with the ohmic component. There were two reasons: first, the appearance of the anomalous plasma resistance near the maximum of the discharge current, where the derivative of the current is small; second, the rather good stability (during the measurement time interval) of the plasma column and of the current channel, which kept the inductance constant. Furthermore, this discharge was distinguished by a large conductivity anomaly ($\sigma_{class}/\sigma_{anom} = 10^3$-$10^4$).

Similar conditions prevail in essentially all high-current straight gas discharges stabilized by a magnetic field. In such discharges it is a simple matter to meet the requirements regarding the critical electric field and — a particularly important point in actual experiments — the critical current density. The primary consideration in planning an experiment is to maintain conditions corresponding to the appearance of collective current interactions in the plasma. In some cases, different results are obtained because of a deviation from these conditions, in experiments which appear similar at first glance. Even in a given experimental device there are factors which change over time (the purity of the vacuum, the surface state of the electrodes and the discharge chamber, etc.) and cause a failure to meet the critical parameters. Appropriate remedial measures can completely restore the effects observed earlier. In most cases a clear indication that the necessary conditions have been arranged is microwave emission from the plasma in the corresponding frequency range; this emission always accompanies the anomalous plasma resistance.

In summary, high-frequency, small-scale instabilities always occur in current-carrying plasmas in strong electric fields and give rise to an anomalous resistance of a collective nature.

Just as the Spitzer resistance of a plasma underlies the method of ohmic heating, the anomalous resistance underlies the method of turbulent plasma heating. Because of the high anomalous resistance, turbulent heating is an effective current heating method, which can be used at high plasma temperatures.

The chapters which follow will contain a detailed examination of the physical nature of the anomalous resistance and of the turbulent heating of plasmas. In the rest of the present chapter we will describe the most important of the experiments which have been carried out on the anomalous resistance and turbulent heating over the past 20 years.

2 EXPERIMENTS ON THE ANOMALOUS RESISTANCE AND TURBULENT HEATING OF PLASMAS

Collective interactions in a current-carrying plasma express fundamental properties of the plasma. It is thus quite natural that certain manifestations of collective interactions in plasmas with currents were observed before a clear picture could be drawn of the anomalous conductivity as a physical phenomenon. Contradictions between experimental data and the theory based on binary collisions were noted frequently, but their meaning was not clear; the cause-and-effect relationships among the various processes had not been established.

In the early days of fusion research, experiments on high-current straight gas discharges revealed neutron, electromagnetic, and x-ray emission from a deuterium plasma even in quite strong stabilizing magnetic fields, in which the heating and acceleration of deuterons by shock compression could be ruled out (see, for example, the COLUMBUS 4 experiment, carried out at Los Alamos[164]). At the time, these effects went unexplained. Looking back at them now, we may assume that turbulent plasma heating was occurring.

In experiments in the ZETA device[32] (UK), an increase in the discharge resistance was observed to accompany an increase in the initial gas pressure. This effect could not be reconciled with the existing understanding; it was suggested that the observed behavior of the resistance was caused by an influx of impurities into the discharge. A careful experiment in which argon was added, however, did not support this suggestion. An unexpected result was obtained on this apparatus in measurements of the ion temperature from the broadening of the spectral lines of impurity ions: The temperature of the impurity ions turned out to depend on the ion charge. A detailed study of this effect was carried out in the first experiments in the AL'FA device[10,11] (Leningrad), all of whose basic parameters were similar to those of the ZETA. These experiments established that the dependence $T_i(z)$ was linear, and they revealed a distortion of the Maxwellian ion distribution in the high-velocity direction. All of these effects can be explained easily in terms of the onset of collective interactions in a current-carrying plasma.

Systematic studies of collective interactions in current-carrying plasmas were begun in the late 1950s and the early 1960s and have continued to the present.

For convenience in reviewing the chronology of the basic experiments on the anomalous resistance and turbulent current heating of plasmas, we have collected all the basic parameters of the experimental devices in Tables 1 and 2.

The anomalous electrical conductivity of a plasma, as a phenomenon associated with a basic parameter reflecting the physics of the process – the electric field in a current-carrying plasma – was first discovered in 1959 in experiments in the GROM-1 device[40] (Khar'kov Physico-technical Institute). It was learned in these experiments that the conductivity of a plasma in a high-current straight discharge stabilized by a strong magnetic field is much lower than the conductivity corresponding to binary

Table 1

Straight discharges

Device, location, year of publication, ref.	l, cm	a, cm	H_{max}, kOe	V_0 kV	n, cm^{-3}	I, kA	nT, eV/cm^3	T_e, keV	T_i, keV
GROM-1, KhPTI, 1959[40]	50	9	0.3	1.5	10^{14}	99	$4\cdot10^{16}$		
Mirror, IAE, 1961[17]		1.8	3		$2\cdot10^{13}$				
GROM-3, KhPTI, 1961[160]	24	4	15	28	10^{15}	200	$3\cdot10^{18}$	3	1
PR-1, IAE, 1964[18]	100	4	1.8	10	$5\cdot10^{12}$	3		0.2	3
TH-4, IAE, 1967[20]	200	3	3.5	140	10^{13}		$5\cdot10^{16}$	5	2
TN-5, IAE, 1968[92]	300	12	12	140	$1.4\cdot10^{13}$		$6\cdot10^{16}$	3	3
Mirror, GGAI Laboratory, San Diego, 1968[239]	180	6	1.6	76	$2\cdot10^{14}$	5	$2\cdot10^{15}$		0.5
NPR-2, IAE, 1969[115]	90	4	25	0	10^{14}		$4\cdot10^{16}$	1	
Mirror, INP, Novosibirsk, 1970[33]	100	3	20	40	$5\cdot10^{13}$		$2.5\cdot10^{15}$	2.5	0.8
Steady-state straight discharge, Oak Ridge 1970[231]	30	0.5	0	9	$3\cdot10^{10}$			0.1	

GROM-4, KhPTI, 1970[136]	30	1.5	50	28	$7\cdot10^{14}$	200	$2\cdot10^{18}$	2	1
BSGII, Nagoya Univ., 1970[257,295]	120	3	2	36	$5\cdot10^{13}$	3	$5\cdot10^{15}$	0.1	
THE OCUII, Osaka, 1970[257]	30	3.5	17	25	$5\cdot10^{14}$				1
Straight rf skin-effect discharge, SRIEPA, Leningrad, 1970[72]	100	2	3	1.5	10^{14}	1	$5\cdot10^{16}$	0.1	0.7
TURHE, Jutphaas, Holland, 1971[245]	80	6	6	40	$5\cdot10^{13}$		$5\cdot10^{17}$	6	2
Mirror, Cornell Univ. 1971[302]	180	4	3	100	$8\cdot10^{13}$	13	$7.5\cdot10^{16}$		1.5
OGRA-2T, IAE, 1971[88]	106	6	15	40	10^{13}	20	$6\cdot0^{16}$		3-10
Explosive discharge of polyethylene filament, NRL, Washington, 1972[285]	3.5	0.1	0	600	10^{19}	1200			0.2
THE MACH II, Yaponiya Univ.,Nagoya 1975[257]	135	2	30	36	10^{14}		$8\cdot10^{17}$	8	2
NPR-2M, IAE, 1977*	10	1	90	45	10^{16}		$2\cdot10^{18}$		

Note: *KhPTI) Khar'kov Physicotechnical Institute, Khar'kov; IAE) Kurchatov Institute of Atomic Energy, Moscow; INP) Institute of Nuclear Physics, Novosibirsk; SRIEP) Efremov Scientific-Research Institute of Electrophysical Apparatus, Leningrad.*

*L. E. Aranchuk, Yu. G. Kalinin, A. S. Kingsep, Yu. I. Shestakov, and V. A. Skoryupin, in: *Thirteenth International Conference on Phenomena in Ionized Gases*, Berlin, DDR, 1977, pp. 855-856.

Table 2

Toroidal Discharges

Device, location, year of publication, reference	R, cm	a, cm	H, kOe	E, V/cm	n, cm^{-3}	I, kA	nT, eV/cm^3	T_e, keV	T_i, keV
VIKhR'-1, IAE, 1964[167]	15	1.5	1	150	10^{13}	2		1	
VIKhR'-2, IAE, 1966 (bumpy torus)[168]	25	2	3	250	10^{13}	1	$7\cdot10^{16}$	0.1	0.01
Plasma betatron, IAE, 1965[157]	20	2	3	250	$5\cdot10^{10}$	1.2		tens of keV	
SIRIUS racetrack stellarator, KhPTI, 1966[43]	600	2.5	20	20	10^{13}		10^{15}	1-9	0.08-0
TWIST stellarator, Culham, 1967[221]	32	5	5	500	10^{13}	6	10^{17}	keV	0.7
TOR-IV rf tokamak, IAE, 1970[5]	18	2	20	250	$3\cdot10^{13}$	5	$3\cdot10^{16}$		
Plasma betatron, Saskatoon Univ. 1970[251,232]	19	2.5	2	100	10^{13}	1	10^{16}	0.3	0.01
TORSO torsatron, Culham, 1972[227]	40	5.5	20	50	10^{13}		10^{16}	10	
TTF tokamak, Frascati, 1974[250,253]	30	4	6	25	$5\cdot10^{13}$	5	10^{15}		
TTT tokamak, Univ. of Texas, 1975[191]	60	10	20	500	10^{13}		$3\cdot10^{16}$	10	0.5
TORTUR tokamak, Jutphaas, 1975[242]	45	9	20	300	10^{14}		$2\cdot10^{17}$		0.6
Tokamak, LPI, 1976[69]	18	5	10	150	$7\cdot10^{13}$		$4\cdot10^{16}$	0.3	

Note: IAE) Kurchatov Institute of Atomic Energy, Moscow; KhPTI) Khar'kov Physicotechnical Institute, Khar'kov; LPI) Lebedev Physics Institute, Moscow.

Coulomb conditions and that it falls off with increasing electric field (Figs. 1 and 5).

In 1961, experiments at the Kurchatov Institute of Atomic Energy, Moscow, revealed, for the first time, a heating of a collisionless plasma as the result of an anomalously rapid absorption of strong magnetosonic waves upon an instability of an electron current flowing across the magnetic field at the wavefront.[17] Measurements showed that the damping length for the magnetosonic wave was on the order of the wavelength and that the dissipation of wave energy in the plasma occurred in a time interval 100 times shorter than the binary-collision time. The temperature of the plasma electrons after the heating in that experiment was measured by a simple calorimetric method and turned out to be on the order of 800 eV at a plasma density of 10^{12} cm^{-3}.

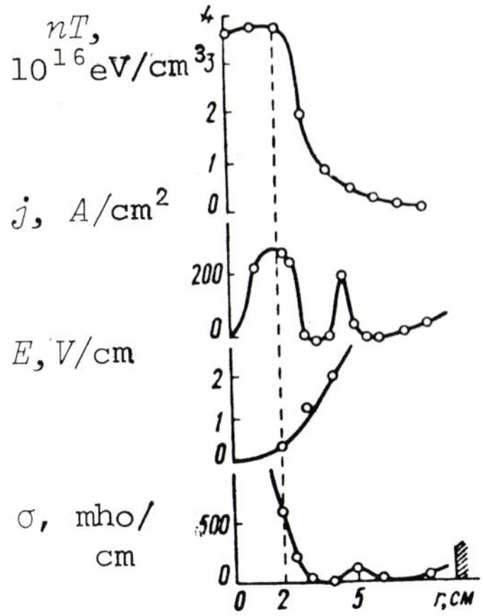

FIGURE 5 Radial profile of the kinetic energy density, the discharge current density, the longitudinal component of the electric field, and the electrical conductivity in the plasma column in the GROM-1 device.[289]

The unambiguous relationship between the onset of the anomalous plasma resistance and the excitation of intense microwaves in the plasma was first established in 1961 in experiments in the GROM-3 device[160] (Khar'kov Physicotechnical Institute) (Fig. 6). These experiments were carried out in a high-current straight gas discharge

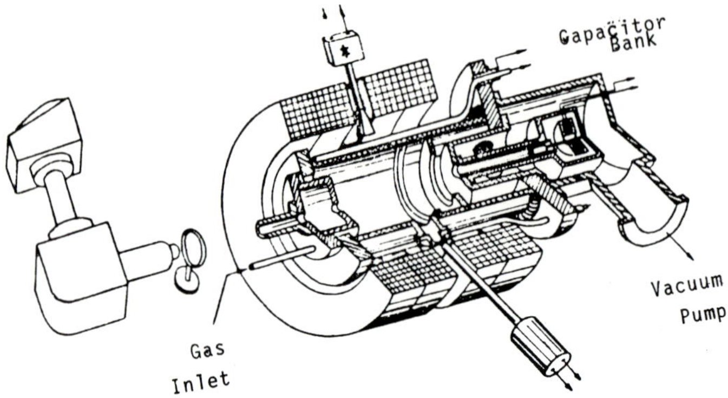

FIGURE 6 The GROM-3 device.

with a high plasma density ($n = 10^{15}$ cm^{-3}), a strong electric field, and a strong stabilizing magnetic field. A system of external antennas detected a superthermal microwave emission from the plasma (over a broad wavelength range) which was correlated with the slowing of the directed stream of plasma electrons and with the appearance of an anomalous resistance (Figs. 2-9). These experiments were the first to reveal a threshold in the dependence of the basic manifestations of collective interactions on the electric field: Below a threshold field, the process did not occur. The anomalous plasma resistance was studied as a function of the electric field at strong fields and also as a function of the working-gas pressure.

In 1964-1966 the turbulent heating of a plasma in a high-current straight discharge was first observed and studied in this device.[162] Analysis of the x-ray emission yielded the electron temperature (on the order of a few keV) and revealed that this temperature was "isotropic" with respect to the direction of the electric field.

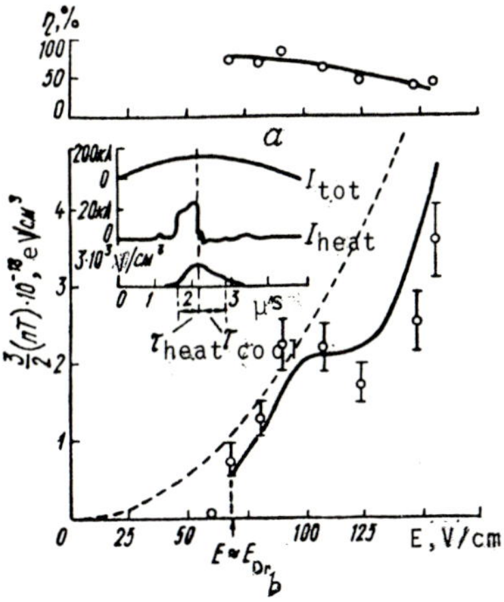

FIGURE 7 Results from the GROM-3 device.[135]
a) Efficiency (the "physical efficiency," η) of the dissipation of the energy supplied to the discharge gap vs. the electric field; *b)* energy of the plasma (nT) vs. the electric field. The dashed curve corresponds to a quadratic law. The oscilloscope traces show the total discharge current (I_{tot}), the current in the axial part of the discharge (I_{heat}), and the diamagnetic signal (nT). The vertical dashed line here is the end of the heating interval (τ_{heat}) and the beginning of the plasma cooling interval (τ_{cool}).

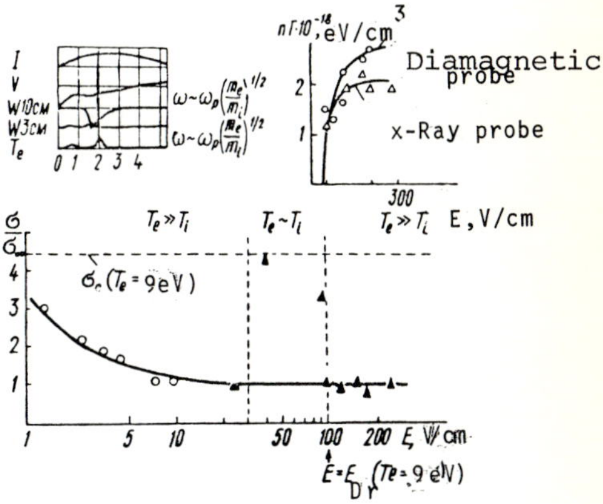

FIGURE 8 Electrical conductivity of the plasma, with the high-field plateau.[289] The oscilloscope traces show, from top to bottom, the total discharge current, the voltage across the discharge gap, the microwave emission from the plasma at a wavelength of 10 cm, the emission at 3 cm, and the signal from an x-ray detector used to measure the electron temperature. o) The GROM-1 device (α = 1.6); ▲) GROM-3 (α = 1.0).

Heating of plasma ions (Figs. 10 and 11) was observed and studied in these experiments[163] (in 1966) under conditions corresponding to the anomalous resistance of a high-current discharge. The ion temperature (up to 2 keV) was measured by a passive particle diagnostic method. Study showed that the ion heating is yet another

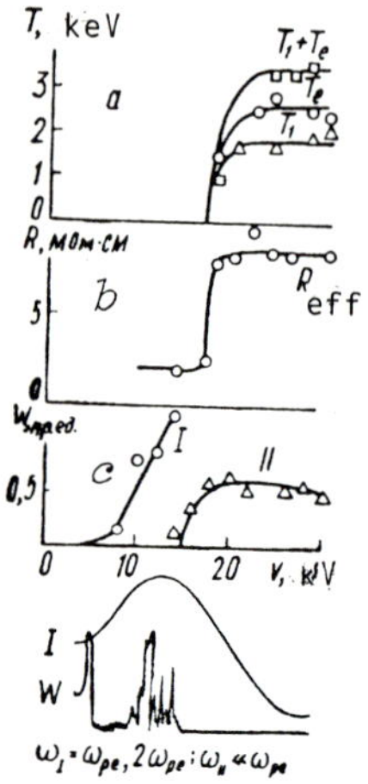

FIGURE 9 Variation of several properties with the voltage to which the capacitor bank is charged.[288] *a*) Electron and ion temperatures; *b*) resistance of the plasma column; *c*) microwave emission from the plasma at 40 GHz (*I* – Initial stage of the discharge; *II* – electron–ion streaming instability). The oscilloscope traces show the total discharge current and the microwave emission.

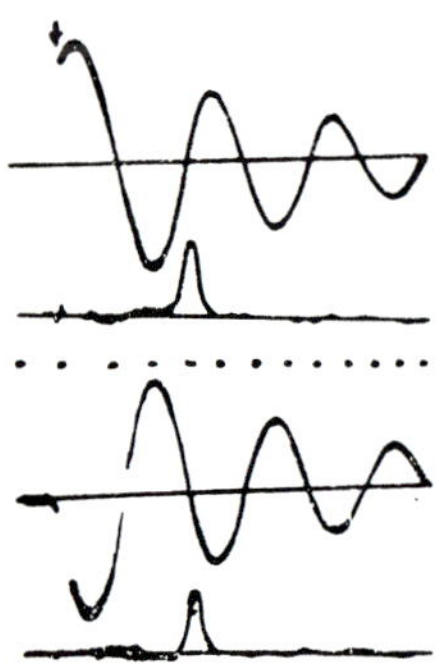

FIGURE 10 Oscilloscope traces of the total discharge current (the sine wave) and of the pulse of fast neutral hydrogen atoms escaping from the discharge (with an average energy of about 1 keV) (the GROM-3 device[163]). *Arrow*) Actual time at which the fast particles appear; *a*) a flux of fast neutrals is observed along the direction of the accelerating electric field; *b*) decelerating field. The time markers are at 1 μs. The time interval on the trace between the arrow and the pulse corresponds to the time required for a particle to cross the diagnostic volume.

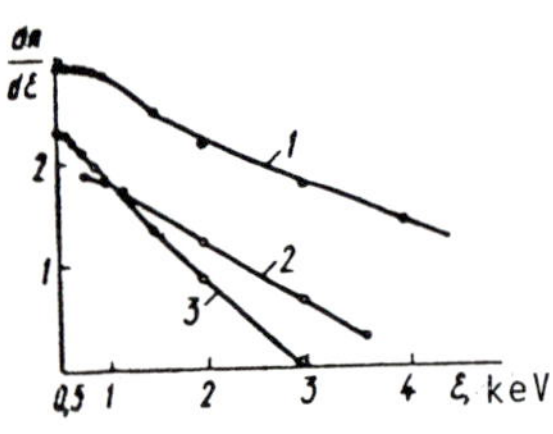

FIGURE 11 Energy distributions of the fast neutral hydrogen atoms escaping from the discharge (the GROM-3 device[289]). *1*) Observation along the magnetic field in the direction of the accelerating electric field ($T_i \sim 2$ keV); *2*) in the transverse direction ($T_i \sim 1.5$ keV); *3*) in the direction of the decelerating electric field ($T_i \sim 1$ keV).

manifestation of collective interactions in a plasma with a current.

Measurements of the plasma diamagnetism and estimates based on measurements with a thermionic probe (with an incandescent platinum foil) yielded temperatures of the same order as those just cited. It was rigorously proved in these experiments that the turbulent heating is a consequence of high-frequency current-driven instabilities in the plasma.

The first experiments on turbulent heating in a toroidal confinement system were carried out in 1964 in the VIKhR'-1 device at the Kurchatov Institute of Atomic Energy, Moscow.[167] An electric field (150 V/cm) far higher than the critical Dreicer field was applied around the torus. The plasma exhibited an anomalously high ohmic resistance under these conditions, corresponding to an effective electron collision rate on the order of 10^9 s^{-1}, far higher than the rate of binary collisions of any type.

An estimate of the electron temperature from the x-ray emission after the heating showed that in a plasma with a density of 10^{12} cm^{-3} the electrons are heated to 10^2-10^3 eV. A spectral study was made of the microwave emission accompanying the turbulent-heating stage, and it was shown that the microwave signal reaches a maximum at frequencies corresponding to the electron plasma frequency.

Further work in the VIKhR'-1 device revealed a Raman (or "combinational") scattering of electromagnetic waves by electron waves in the turbulent plasma at a frequency ω_{pe} ; this effect was exploited to estimate the degree of turbulence of the plasma of the turbulent current discharge: $\frac{W_e}{nT} \approx 10^{-1}$.

Turbulent heating in a magnetic mirror was first achieved in 1964 at the Kurchatov Institute (in the PR-1 device[18]). A current flowing along the magnetic field between electrodes heated the plasma, which was produced beforehand by plasma injectors at the magnetic mirrors; these injectors doubled as end electrodes. These experiments revealed an anomalously high resistance and a pronounced heating of the plasma electrons. When adiabatic compression was also used, a plasma with a density above 10^{12} cm^{-3} was produced. Estimates based on the neutron emission yielded an ion temperature of about 3keV. The hot plasma was confined in a mirror system for a time on the order of 2 ms.

Turbulent plasma heating in a toroidal confinement system with a bumpy magnetic field[168] was achieved in experiments in the VIKhR'-2 device (at the Kurchatov Institute). In these experiments an electric field (E_{max} = 250 V/cm) far stronger than the Dreicer field was applied around a plasma loop produced by titanium injectors (n=10^{11}-10^{13} cm^{-3}). The plasma confinement in the system was studied after a brief current pulse (lasting about 1 μs).

These experiments revealed a pronounced anomaly in the plasma conductivity, accompanied by intense superthermal microwave emission at the frequencies ω_{pe} and $2\omega_{pe}$. Two characteristic plateaus were found on the descending curve of the conductivity vs. the electric field. Data from diamagnetic measurements during the current flow showed that the plasma was heated ($nT \approx 7 \cdot 10^{15}$ eV/cm^3) to a temperature of a few hundred electron volts and that after the current pulse the energy in the plasma decayed

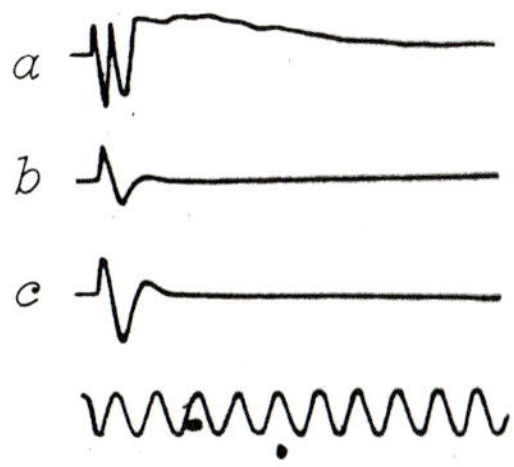

FIGURE 12 Oscilloscope traces of the diamagnetic signal (*a*), of the loop voltage (*b*), and of the current (*c*) in the plasma in the VIKhR'-2 device.[169] The frequency of the calibration signal at the bottom is 1 MHz.

slowly, in accordance with the confining properties of the particular device (Fig. 12). These experiments showed that it is possible in principle to confine a turbulently heated plasma in a toroidal confinement system after the end of the heating-current pulse.

Turbulent heating in a magnetic confinement system of the stellarator type was first achieved in 1965 (in the SIRIUS racetrack stellarator at the Khar'kov Physicotechnical Institute[43,45]; Fig. 13), under conditions with prolonged plasma confinement in a closed system.

It follows from the results obtained in the experiments in this device (Fig. 14) that at electric fields below the critical Dreicer field the current in the plasma is sinusoidal. When the field is raised above the Dreicer field the current flows only during the first voltage half-period and then drops essentially to zero. At this time the plasma is the source of intense x-ray and microwave emission over a broad frequency range.

Measurements of the resistance of the plasma column at various ratios of the applied electric field to the critical Dreicer field, E/E_{Dr}, showed that the resistance is anomalously high over the entire field range studied, $0.2E_{\mathrm{Dr}} \lesssim E \lesssim 10E_{\mathrm{Dr}}$. However, while the resistance varies only slightly at $\frac{E}{E_{\mathrm{Dr}}} < 2$, it increases sharply at

FIGURE 13 Photograph of the SIRIUS stellarator.

$E > 2E_{Dr}$. It was learned that this anomalous resistance depends on the degree of magnetization of the plasma and also on the ratio of the electron cyclotron frequency to the electron plasma frequency.

According to x-ray measurements, a substantial fraction of the plasma electrons acquire an energy corresponding to temperatures between 1 and 9 keV.

Experiments on turbulent current heating of plasmas with densities of 10^{12}-10^{13} cm^{-3} in strong electric fields (100-500 V/cm) were carried out in the TWIST toroidal stellarator (Culham, UK[221]) in 1966-1967. The plasma was observed to have a highly anomalous conductivity (Figs. 15 and 16). The plot of the conductivity against the electric field was stepped, with plateaus of progressively lower conductivity appearing as the field was increased. An effective heating of electrons (to 2-3 keV) and ions

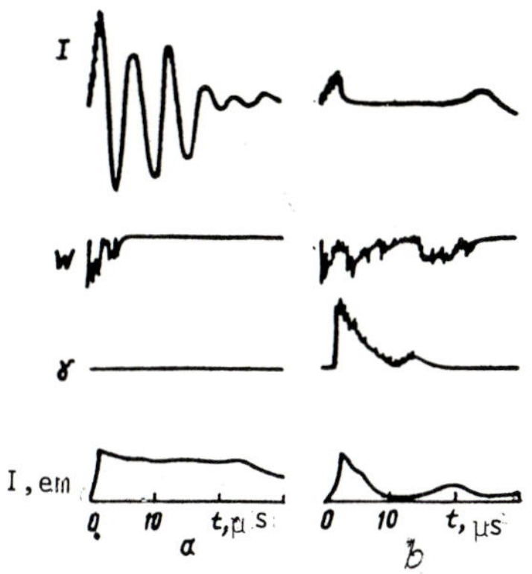

FIGURE 14 Oscilloscope traces of the total discharge current, of the microwave emission from the plasma, of the signal from the x-ray detector used to measure the heating of the electrons, and the integrated emission from the plasma in the SIRIUS stellarator.[43] *a*) Weak electric field ($E < E_{Dr}$), with collective interactions only poorly expressed; *b*) strong electric field ($E > E_{Dr}$), with a highly anomalous plasma resistance.

(to 0.7 keV) occurred according to x-ray, diamagnetic, and passive particle diagnostics.

The rf fluctuations and the microwave emission from the plasma were measured in these experiments, and the ion distribution function was found to have two peaks. The Buneman instability was observed experimentally for the first time, under certain conditions.

The work begun with the TWIST stellarator was continued with the TORSO torsatron.[227] For this magnetic confinement system, of the stellarator type, the wall of

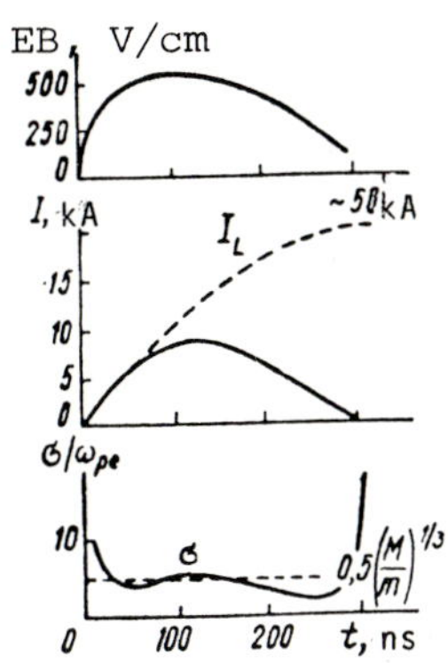

FIGURE 15 Typical time dependence of the discharge parameters in the TWIST device[221] under anomalous-resistance conditions. From top to bottom: the torus loop voltage E, the discharge current I (I_L is the inductive component of the current), and the normalized plasma conductivity σ/ω_{pe}.

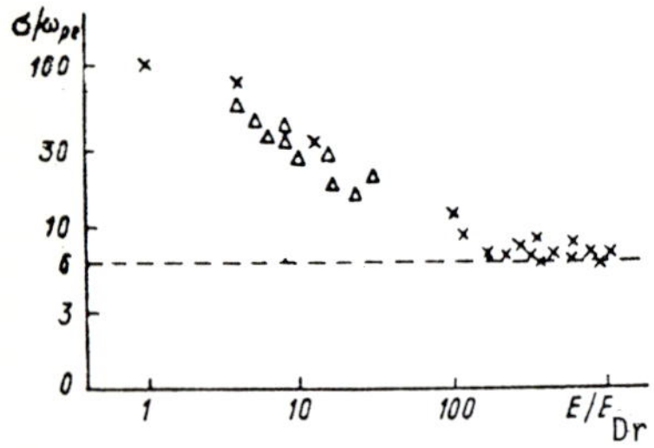

FIGURE 16 Normalized plasma conductivity vs. the electric field (expressed in units of the Dreicer field E_{Dr}). x) TWIST; Δ) TORSO.[221,227]

the vacuum chamber is far from the plasma (Fig. 17; the magnetic-field winding is inside the vacuum chamber), so that it is difficult for impurities to reach the plasma from the wall. Turbulent heating is achieved by passing a short current pulse (with a rise time on the order of 3 μs) through a plasma produced beforehand. The maximum electric field in these experiments was 80 V/cm. During the current flow in the plasma, rf instabilities gave rise to an anomalous resistance and heated the electrons to several keV.

The largest mirror systems, the TN-4 and TN-5 (at the Kurchatov Institute),[20,92] were constructed in 1967-1968. These devices were intended specifically for a study of turbulent heating in straight systems (Fig. 18). In a typical experiment, titanium injectors at the magnetic mirrors produced a column of a highly ionized plasma which was insulated from the chamber wall. The injectors doubled as electrodes for producing a current to flow along the magnetic field. By adjusting the delay of the discharge current with respect to the operation of the injectors it was possible to arrange various plasma conditions in the system, and as a result a variety of current-driven instabilities occurred.

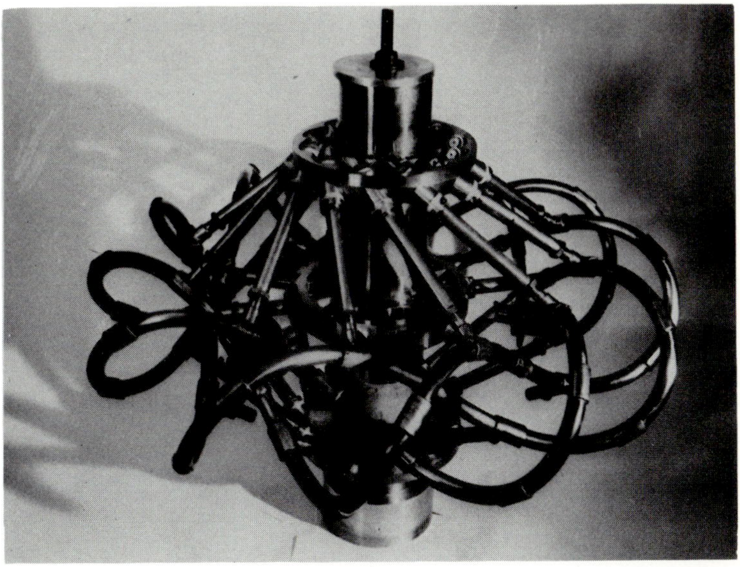

FIGURE 17 Magnet system of the TORSO torsatron with turbulent plasma heating.

A very rapid heating of a plasma with a density $2 \cdot 10^{13}$ cm^{-3} was achieved in the TN-4. The heating time (0.5 μs) corresponded approximately to the time at which the discharge current reached its maximum. In various experiments, turbulent heating was studied under conditions such that the plasma pressure attained was lower than, equal to, and higher than the maximum magnetic pressure at the periphery of the plasma column. The physical efficiency of the turbulent heating was determined and found to reach a maximum of about 2.5%. It was learned that the plasma heating increases with increasing magnetic field of the confinement system, reaching saturation at a high field (10 kOe). Heating of the ions was studied, and it

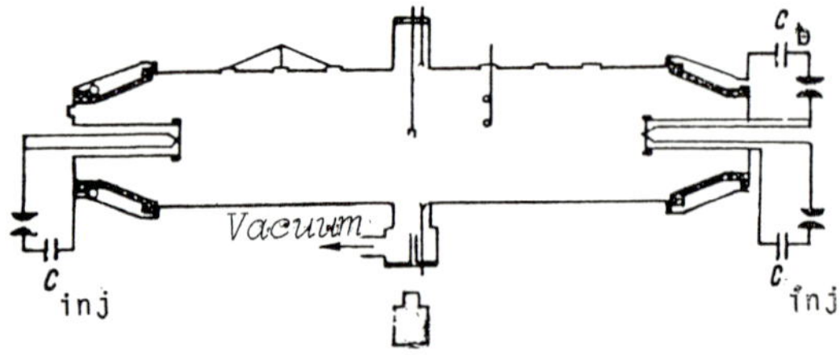

FIGURE 18 The TN-5 mirror system.

was demonstrated that conditions corresponding to the occurrence of an ion acoustic instability could be arranged.

In experiments in the TN-5 device, turbulent plasma heating was achieved in a record large plasma volume (more than 100 liters) under conditions corresponding to the flow of the heating current in a skin layer (Fig. 19). At a plasma number density of $2 \cdot 10^{13}$ cm^{-3} and an energy density on the order of $6 \cdot 10^{16}$ eV/cm^3, the average ion energy reached about 3 keV. The ion distribution function measured over the energy range 0.5-20 keV did not have a maximum; the density of ions with energies in this range amounted to about a third of the total plasma density. It was found that the efficiency of the turbulent ion heating was high and that during the stage of plasma confinement in this mirror system the ions carried most of the energy. It was demonstrated that the current-carrying plasma was homogeneous

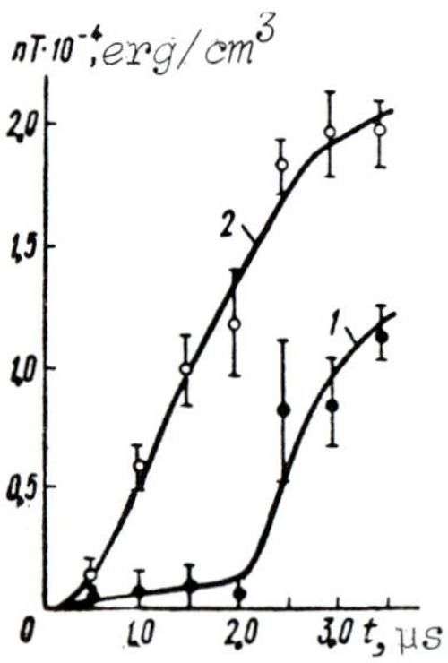

FIGURE 19 Time dependence of the energy in the plasma in the TN-5. *1)* Axial region; *2)* in the skin layer.[6,92]

along the axis of the device and that the heating was of a volume nature.

It was also shown that the plasma heating initially occurs in a thin shell at the periphery of the plasma column, and later in the central part. This effect was subsequently explained[6] on the basis of experimental data as a consequence of an ion acoustic instability in a two-stream plasma; the instability caused an effective deceleration of the plasma streams coming from the injectors. The directed-motion energy of the streams was converted into thermal energy, and the heating of the plasma column propagated radially inward from the skin layer toward the axis at the ion acoustic velocity. This heating was accompanied by intense noise at frequencies on the order of the ion plasma frequency. Experiments carried out in later years showed that a turbulent discharge is an effective tool for converting the kinetic energy of oppositely directed plasma streams into thermal energy and thus represents a promising method for producing hot plasmas in large-scale devices.

An extensive program of research on turbulent current heating was carried out in the large mirror system of the

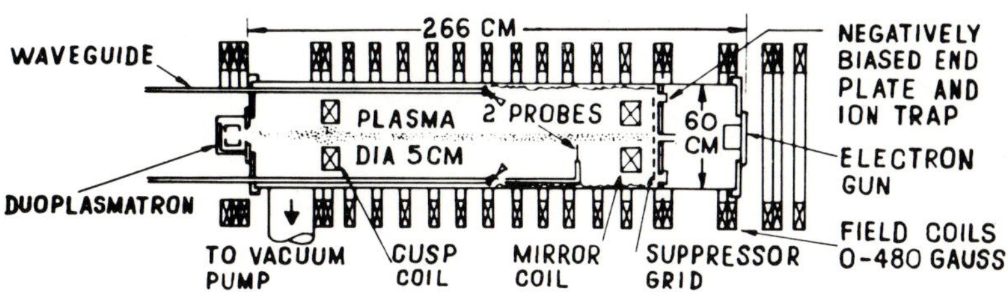

FIGURE 20 Turbulent-heating mirror system.[302]

Gulf General Atomic plasma laboratory in San Diego in 1968 (Ref. 239). The apparatus was later moved to Cornell University, where another research team[302] continued the work (Figs. 20 and 21).

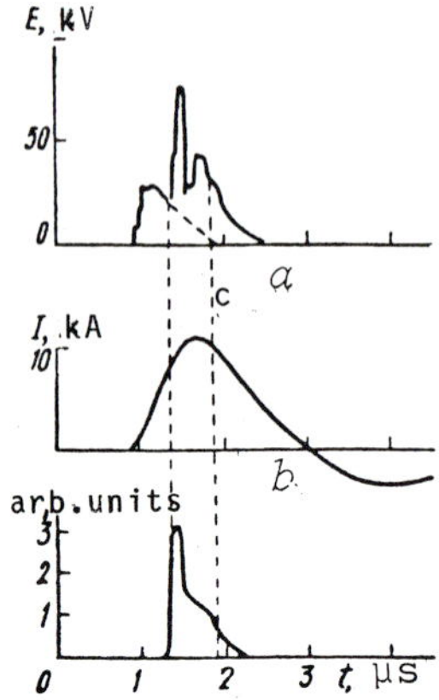

FIGURE 21 *a)* Voltage across the discharge gap; *b)* total current; *c)* signal from an x-ray detector used to measure the electron temperature (from the mirror system at Cornell University).[302]

Experiments carried out in this device showed that it is possible to heat a plasma in a large magnetic mirror system by passing a current along its axis. The efficiency of the energy transfer from the capacitor bank to the plasma (the technical efficiency) was found to be 3-4%. The energy transferred to the plasma was found to vary quadratically with the amplitude of the heating current. Heating of the plasma ions was observed; the ion energy distribution was found to be a "two-temperature" distribution with peaks at 500 and 1500 eV. Diamagnetic measurements, which yielded the same total temperature, showed that it was primarily the ions which were heated. Only a few hot electrons, with energies above 5 keV, were observed.

Heating was observed during a pronounced anomaly in the plasma conductivity. Magnetic-probe measurements revealed that the discharge current flowed in a skin layer. The dynamics of this layer was studied.

A thorough study of the physics of the turbulent heating of a current-carrying plasma was carried out in 1971 in the NPR-2 mirror device[115,93] (Kurchatov Institute) It was shown that turbulent heating by a longitudinal current, which results from the dissipation of the energy from an external source across the anomalous resistance of the plasma column, is an efficient method for heating plasmas (the efficiency is 25%). The ratio u/v_S was found to be 2-8, and the anomalous resistance was found to depend on the plasma density, the electric field, and the magnetic field in the manner predicted by the theory for the ion acoustic instability. According to the experimental results, the energy loss during the heating occurs primarily along the magnetic field, to the anode of the straight discharge, where 70-85% of the energy supplied to the discharge gap is lost. This effect was ascribed[115,93] to a current-convective instability in the anomalous-resistance plasma. Direct measurements showed that the discharge current did not flow in a skin layer, that the electric field was distributed uniformly over the length of the discharge, and that the potential drop near the cathode was very small, no more than $T_e/e \sim$ 500-1000 V at a voltage on the order of 30 kV across the gap. It was concluded on this basis that the anomalous resistance was a volume effect. Measurements of the x-ray emission from the anode of the straight discharge in the NPR-2 showed that the plasma heating in these experiments could not be ascribed to a beam instability. A broadening of the Hα and Hβ hydrogen spectral lines was observed concurrently with the anomalous resistance and used to estimate the electric fields of the ion acoustic waves (30-50 kV/cm). The distribution of these fields turned out to be very anisotropic according to a polarization effect in the Stark profile of the Hα line.

In a later modification of this apparatus (the NPR-2M, in 1977; see the footnote in Table 1), turbulent heating was used to produce a plasma with a density of 10^{16} cm^{-3} containing an energy of $2 \cdot 10^{18}$ eV/cm^3.

Turbulent heating of a dense plasma ($n = 10^{15}$ cm^{-3}) was achieved in a mirror system with a strong magnetic field (50 kOe at the mirrors; mirror ratio of 1:2) and with a strong electric field (E = 200 V/cm) in experiments in the GROM-4 device (Fig. 22) in 1970-1971 (Khar'kov Physicotechnical Institute).[136] These experiments showed that in a sufficiently strong electric field a series of

FIGURE 22 Photograph of the GROM-4 mirror system with turbulent heating of a dense plasma.

current-driven instabilities occur in the discharge and give rise to an anomalously low plasma conductivity, superthermal microwave emission in the frequency range from ω_{pe} to ω_{pi} and below, and an effective heating of the plasma.

According to diamagnetic measurements (Fig. 23), the plasma energy increases rapidly during the current flow, to a value approaching 10^{19} eV/cm^3. After the stage of the current flow, a plasma with a thermal energy density on the order of 10^{18} eV/cm^3 is confined in the mirror system for a time corresponding to the theoretical plasma lifetime in the system. The plasma energy increases in proportion to the square of the voltage applied to the discharge gap.

In experiments with a deuterium plasma, neutron emission was achieved; estimates based on the intensity of this emission put the ion temperature at about 1 keV.

The net result of the experiments carried out in this device was to demonstrate that turbulent heating in a magnetic mirror system could produce a plasma with a density n and a temperature T close to the values required for fusion and that this plasma could be confined for a significant time, limited only by the properties of the confinement system itself.

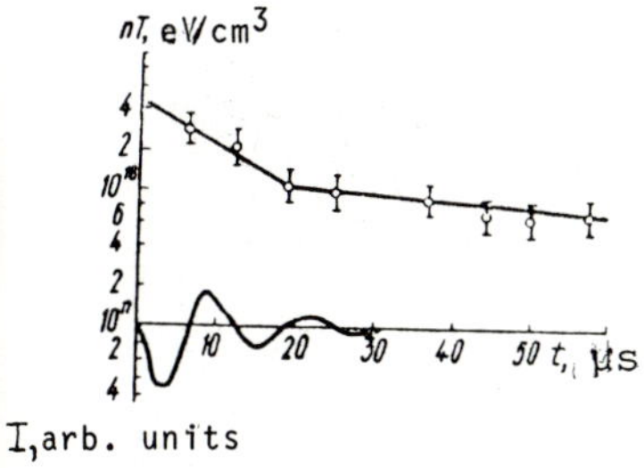

FIGURE 23 Time dependence of the plasma pressure and of the total discharge current in the GROM-4 device.[136]

In 1971, turbulent plasma heating was used to produce a hot plasma (suitable for study of plasma stability in a magnetic mirror system) in the well-known OGRA-2T device[88] (Kurchatov Institute). During the flow of a longitudinal current, two stages with an anomalously high plasma resistance were observed in the evolution of the discharge (Fig. 24). In the first stage, during the current rise, the plasma ions were heated to temperatures at the keV level, but no high-energy electrons appeared; in the second stage there was an abrupt cutoff of the current, and an intense stream of fast electrons formed, with an energy on the order of the potential difference applied to the plasma. The first stage was interpreted as the onset of an ion acoustic instability, while the second was ascribed to beam-plasma interactions. The total energy dissipation was roughly the same in the two stages.

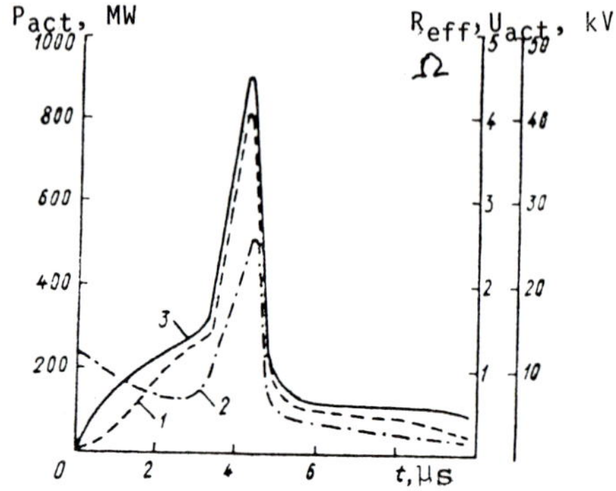

FIGURE 24 Time dependence of the active power (1), of the active plasma resistance (2), and of the active discharge voltage (3) in the OGRA-2T device in the turbulent heating regime.[88]

The flow of discharge current was accompanied by a rapid expansion of the plasma, during which about half of the plasma particles escaped confinement, reducing the overall density by about an order of magnitude. This effect was probably a consequence of a disruption of the equilibrium between the plasma and the magnetic field.

During the first stage of the heating in these experiments, the plasma density was 10^{13} cm^{-3}, and the average ion energy was 2-3 keV. After the current cutoff, with an expanding plasma remaining in the system, the average ion energy was on the order of 10 keV.

Further experiments by Zhil'tsov et al.[88] showed that the radial expansion of the plasma could largely be avoided by increasing the magnetic field and reducing the input energy.

The turbulent heating of a plasma in a straight rf discharge with a current skin effect and with a magnetic field was studied at the Efremov Scientific-Research Institute of Electrophysical Apparatus,[82,35] Leningrad, in 1970. In these experiments a rf current pulse ($\omega \doteq 10^7$ s^{-1}, $I = 1$ kA) from a pulse circuit was passed through the plasma. In the skin layer in this discharge, rf instabilities occur and give rise to an anomalous discharge resistance and to a pronounced heating of the electrons and ions. These experiments showed for the first time that the ion energy distribution consists of a main group of particles with $T_i = 50$ eV and a small group with $T_i \cong 700$ eV; the average electron temperature in the plasma is on the order of 100 eV. Each of the ion groups can be described by a corresponding Maxwellian energy distribution. This effect (a two-temperature distribution) was later observed in many other experiments on turbulent plasma heating.

The dynamics of a turbulent current discharge in a mirror system was studied in 1969 at the Institute of Nuclear Physics, Novosibirsk,[33] where nanosecond electronic techniques and novel diagnostic apparatus made it possible to determine the time evolution of the basic processes during a single discharge pulse. It was shown that when the current reaches a critical level the plasma conductivity becomes anomalously low because of an instability of the Buneman type. The instability is accompanied by microwave emission from the plasma at $\sim\omega_{pe}$ and by a noisy modulation of the discharge current at $\omega \geq \omega_{pi}$. A pronounced heating of the plasma, to a few keV, is observed.

The electron temperature was measured in these experiments by several methods: by an x-ray method, with a multigrid probe, and by a diamagnetic method. The results obtained for the "transverse" temperature by the different methods agreed. The results obtained by the x-ray and multigrid-probe methods put the "longitudinal" temperature at T_e = 2-3 keV in the electron-acceleration direction and at 1.5-2 keV in the deceleration direction (at the cathode). The "transverse" temperature was about 2.5 keV.

Turbulent heating of the plasma ions was observed in the experiments; the ion temperature was determined with a multichannel electrostatic analyzer. The ions were found to have a two-temperature energy distribution, with the hotter group of ions (with T_i = 0.8 keV) constituting about 25% of the total density. The sequence in which the electrons and ions were heated was determined; it turned out that the electrons are heated first. They acquire a temperature of 2.5 keV in 100 ns, while the ions reach 0.8 keV in 0.4-0.5 μs.

The dynamics of the potential distribution along the discharge gap was studied in those experiments with an array of electrostatic Langmuir probes. The measurements showed (Figs. 25 and 26) that when the anomalous plasma resistance begins to appear (about 0.4 μs after the beginning of the discharge current) the active component of the voltage, which is a result of an instability, is distributed nearly uniformly over the length of the plasma column. At the maximum of the discharge current (at about 0.6 μs) there is a significant potential drop near the cathode, while after the maximum (at about 0.8 μs) the voltage drop

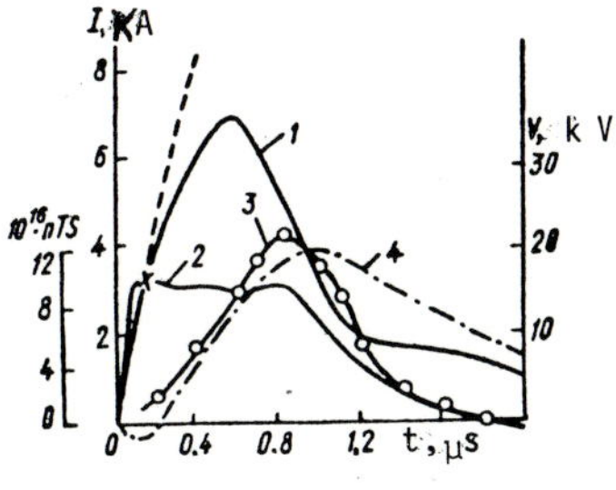

FIGURE 25 Time dependence of various characteristics of the straight discharge (at the Institute of Nuclear Physics, Novosibirsk.[33] *1)* Discharge current; *2,3)* the total voltage and its active component, respectively, on the plasma column; *4)* the diamagnetic signal, nTs.

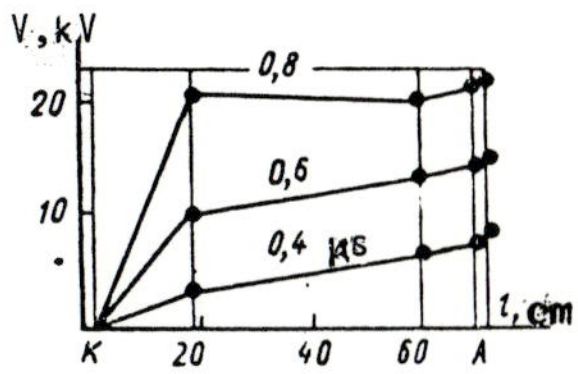

FIGURE 26 Potential distribution along the discharge at various times in the high-current straight discharge at the Institute of Nuclear Physics, Novoskibirsk.[33]

occurs primarily over a region 20 cm long (slightly more than a quarter of the entire gap length) near the cathode.

These experiments included the most comprehensive study to date of the dynamics of the turbulent heating in a straight discharge.

An experiment carried out in 1970 at Oak Ridge[231] with a steady-state high-voltage discharge was extremely interesting in connection with the manifestations of the anomalous plasma conductivity in very different devices. The experimental dependence of the steady-state anomalous plasma resistance on the applied electric field was obtained over a very broad field range (from 0.06 to 300 V/cm) for a plasma density $n = 3 \cdot 10^{10}\ \mathrm{cm}^{-3}$. There are four regions on this dependence: a high plateau (at weak fields), a region in which the conductivity falls off in proportion to $1/E$, a region with a decay in proportion to $1/\sqrt{E}$, and then another, lower, plateau (at strong fields). Having measured the drift and thermal velocities of the electrons as well as the intensity of the ion acoustic turbulent fluctuations, Hirose et al.[231] interpreted the decaying regions on this curve as regions of an ion acoustic instability, while the high-field plateau corresponds to the onset of a Buneman instability, with intense rf oscillations.

An interesting experiment on heating in an extremely dense plasma ($n = 10^{19}$ cm^{-3}) in a straight discharge was carried out in 1972 at the Naval Research Laboratory in Washington, DC.[285] A polyethylene filament 10-190 μm thick and 3.5 cm long was exploded by a current of 1.2 MA. The length of the current pulse was $5 \cdot 10^{-8}$ s, and the voltage between the electrodes was 600 kV. The plasma column in the discharge had a diameter of 1-2 mm (according to measurements with an x-ray pinhole camera). A pronounced ion heating was inferred from the neutron emission in experiments with deuterated polyethylene (with a 99% enrichment in D).

An extensive research program on turbulent plasma heating was carried out at the Institute of Plasma Physics at Jutphaas, Holland, with the TURHE device[245] (1971) and the TORTUR device[242] (1974-1975).

Turbulent current heating of a dense plasma (10^{13}-10^{14} cm^{-3}) was carried out in the TURHE mirror system with a strong longitudinal electric field (E_{max} = 600 V/cm); the plasma was produced beforehand in the system by an arc. A highly anomalous plasma conductivity was observed, and the decreasing dependence (with plateaus) of the conductivity on the electric field which had been reported previously was confirmed. The plasma electrons were heated to maximum temperatures $T_{e\,max} \sim 8$ keV, while the ions had two-temperature energy distributions with $T_{i1} \approx 0.75$ and $T_{i2} \approx 6$ keV. The plasma energy reached $5 \cdot 10^{16}$ eV/cm^3.

A study of the rf characteristics of the plasma revealed three stages in the discharge. In the first, electromagnetic radiation was emitted from the plasma at frequencies below ω_{pe}, and there was intense electron heating. In the second stage, there were ion acoustic and plasma waves, which were emitted at high frequencies up to $10\omega_{pe}$; the ions were heated. In the third stage the emission spectrum consisted of discrete peaks at $n\omega_{pe}$, caused by a strong Langmuir turbulence, while ion acoustic waves were not excited.

In the TORTUR tokamak turbulent heating was used to produce a hot, dense plasma (10^{13}-10^{15} cm^{-3}). A current was passed through a plasma with a density above the critical value set by the ion acoustic velocity. The electric field reached 200 V/cm and was far higher than the Dreicer field. A highly anomalous plasma conductivity was observed. The effective collision rate inferred from measurements of σ_{anom} was $\nu_{eff} \approx 0.5\omega_{pi}$. The electromagnetic emission observed during the existence of the anomalous plasma resistance was at a frequency near ω_{pi}. The electrons and ions were heated intensely to temperatures on the order of 1 keV. The electron temperature was determined from the Thomson scattering of a laser beam, while the ion temperature was determined from the fast neutrals escaping from the discharge. The plasma energy was determined from the diamagnetic signal. The results found by these mthods agreed well.

It was concluded from these experiments that an ion acoustic turbulence causes an effective heating of electrons and ions in a dense plasma ($n \approx 10^{14}$ cm^{-3}) in a tokamak. No serious loss of plasma occurs during the heating stage itself, while subsequent confinement of the heated plasma requires short-circuiting the current in the plasma.

Some interesting research on turbulent current heating was carried out with the TOR-IV tokamak[5] at the Khurchatov Institute in 1970. Some earlier modifications of this apparatus had demonstrated that turbulent heating could be arranged in a torus through the excitation of large-amplitude oblique magnetosonic waves ($nT = 5 \cdot 10^{15}$ eV/cm^3; the efficiency at which the wave energy absorbed by the plasma was converted into heat reached 40%). Experiments were subsequently carried out on the heating of a plasma by a direct current flowing along the toroidal tokamak magnetic field. A large-amplitude, 1-MHz electric field was applied around the plasma loop from a pulse circuit as a pulse of an rf field ($\tilde{E}_{max}$ = 150 V/cm) or as a unipolar field ($\tilde{E}_{max} \approx$ 40 V/cm). The plasma ($n = 5 \cdot 10^{13}$ cm^{-3}) was produced and heated beforehand by an approximately steady current. These measurements demonstrated that the heating efficiency was quite high in both the oscillatory case ($nT = 3 \cdot 10^{16}$ eV/cm^3) and the case of a unipolar current pulse with a steep front ($nT = 1.2 \cdot 10^{16}$ eV/cm^3). A novel diagnostic method used in these experiments made it possible to determine the average electric fields of the pulsations and the degree of plasma turbulence (see Section 4 of Chapter VI).

Turbulent plasma heating in a toroidal tokamak betatron was studied in 1970-1976 in a laboratory in Saskatoon

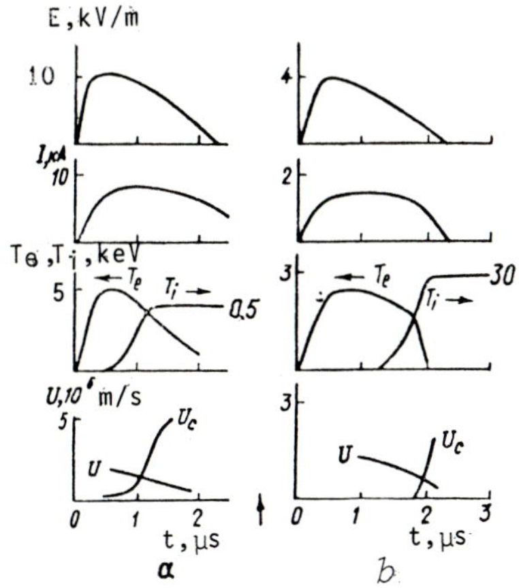

FIGURE 27 Time dependence of the electric field, of the discharge current in the plasma, of the electron and ion temperatures, of the electron drift velocity, and of the ion acoustic velocity.[233] *a)* Discharge in helium; *b)* in argon.

(Saskatchewan).[251,233] A strong electric field (E_{max} = 120 V/cm) was applied in a short pulse (~2 μs) to a helium plasma (and also an argon plasma) produced beforehand by an rf or low-frequency discharge. An anomalous resistance appeared during the current flow in the plasma (Fig. 27), accompanied by a high level of plasma fluctuations, microwave emission, and x-ray emission. Heating of electrons and ions was observed. According to the diamagnetic measurements and measurements with an electron energy analyzer, the electrons in helium were heated to 5 keV and those in argon to 3 keV. According to spectroscopic measurements, the ions were heated to $T_{i max} \approx$ 300 eV in helium and $T_{i max} \approx$ 30 eV in argon. An ion acoustic instability was found to be responsible for the ion heating under those experimental conditions.

Some experiments on overcoming the skin effect were carried out in this apparatus in 1976 (Ref. 189). These experiments were important for the use of turbulent heating in large systems. It was found that an alternating electric field around the loop can "mix" the skin layers and cause a heating of the central part of the plasma. The experiments revealed a heating over the entire plasma cross section (T_e = 2-3 keV, $T_i \cong 0.1$ keV, $n = 10^{13}$ cm^{-3}) in loop fields with E_{max} = 30 V/cm and a frequency of 60 kHz.

Since 1970 a team of Japanese physicists have been carrying out serious research on the mechanism for turbulent heating in several straight devices[257]: BSG II, THE MACH II, and OCU II. These experiments have not been simple repetitions of previous experiments at higher levels. Instead, state-of-the-art high-speed diagnostic techniques have revealed new results, have made it possible to follow the dynamics of the events in the course of a single discharge, and have made it possible to refine earlier results, most of which were obtained by averaging over a large number of discharges (Figs. 28-30).

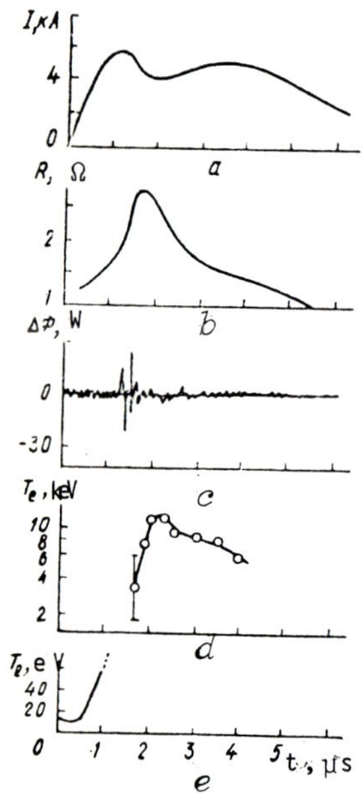

FIGURE 28 Time dependence of the total discharge current (a), of the discharge resistance (b), of the low-frequency fluctuations (c), and of the electron temperatures (d—according to x-ray measurements) in the THE MACH II turbulent-heating mirror.[257]

Experiments in the BSG II, the largest of these three devices, showed that an anomalously high resistance (a resistive surge) appears as the current flows in the plasma. This surge is accompanied by intense fluctuations of the plasma near ω_{pi}, by low-frequency activity in the plasma, and by a heating of ions (to about 20 eV) and electrons (to a temperature on the order of 100 eV). Ions of different species (argon and neon) in the discharge are heated nearly simultan-

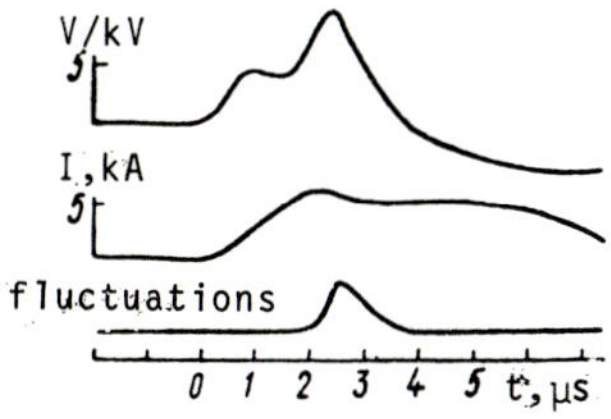

FIGURE 29 Oscilloscope traces of the voltage across the electrodes, of the total discharge current, and of the low-frequency fluctuations in the plasma in the BSG II device.[291]

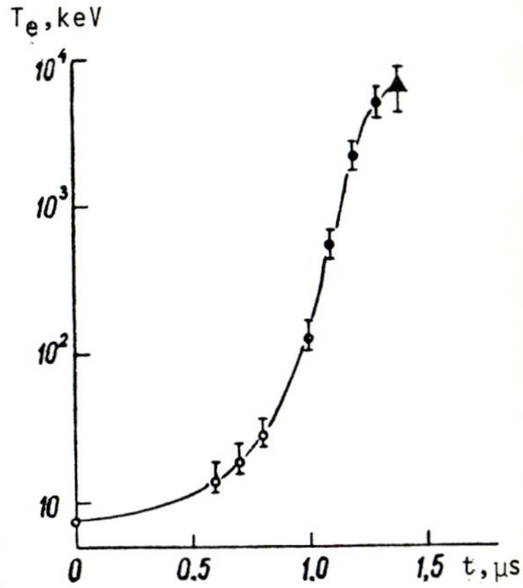

FIGURE 30 Time dependence of the electron temperature in the THE MACH II device.[188,238] o) From laser-scattering measurements; ●) from microwave-scattering measurements; ▲) from measurements of the soft x-ray emission of the plasma.

taneously to an identical temperature.

The most interesting results were obtained in the THE MACH II, which had the strongest magnetic field. As in the BSG II, the resistive surge or anomalous resistance was observed; now it was even more clearly expressed. The surge was accompanied by the onset of a high level of plasma fluctuations, detected by a floating double probe. The fields of the rf microscopic fluctuations in the plasma (at frequencies $\sim\omega_{pe}$) which appear during the resistive surge were determined from observations of the satellites of forbidden spectral lines of helium. These fields turned out to be on the order of several kilovolts per centimeter. Pronounced heating of electrons and ions was observed during the existence of the anomalously high plasma resistance. The electron temperature (see Figs. 28 and 30) was determined by laser scattering ($T_{e\max} \approx 70$ eV) before the resistive surge; then it was determined from the x-ray emission from a graphite target ($T_e \approx 10$ keV). The investigators ascribed the disappearance of the laser-scattering signal upon the onset of turbulence to a deformation of the initially Maxwellian distribution function into a square distribution function and to a high level of parasitic emission from the plasma. The ion temperature (2 keV) was determined by analyzing the fast neutrals escaping from the discharge; the average ion temperature was found to depend linearly on the ion mass; this dependence was found for only the hot component of the ion energy distribution. The dynamics of the increase in the ion temperature during the current pulse was demonstrated.

In the OCU II it was primarily ions which were heated (to 1-4 keV). Several bursts of ion heating were observed during the current pulse, and oscillations were observed at frequencies near ω_{ci}. The ion energy distributions were clearly two-temperature distributions.

Turbulent heating in a large tokamak was studied in 1971-1975 in the plasma laboratory of the University of Texas, Austin.[191] In this device, which was designed especially for turbulent heating, a short pulse (~1 μs) of a strong electric field ($E \gtrsim 200$ V/cm; 100 times the Dreicer field) was applied around a cold plasma column ($T_e \approx 5$ eV) produced beforehand. Intense rf oscillations were excited during the flow of the fast current in the plasma, and these oscillations gave rise to a superthermal microwave emission from the discharge and to an anomalously high plasma resistance. A pronounced heating of plasma electrons (to about 10 keV) and of ions was detected from the x-ray emission from a graphite target and by passive particle diagnostics, respectively. The ions had a two-temperature energy distribution (Fig. 31), with a low-energy part stretching up to 1 keV and a tail at about 5 keV.

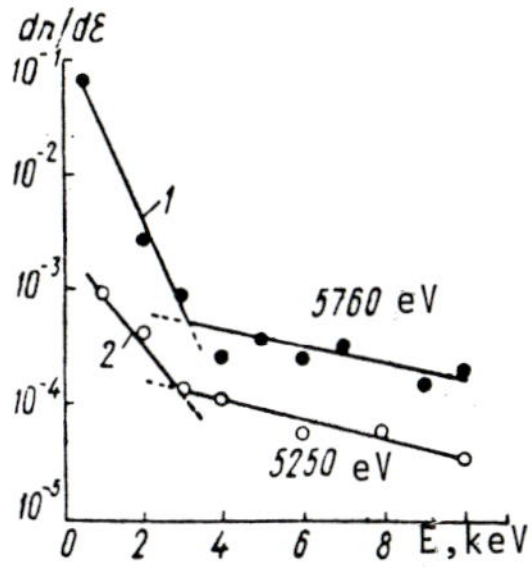

FIGURE 31 Energy distribution of the fast neutrals escaping from the discharge in the TTT device[191] for two plasma densities: *1)* $n = 2\cdot10^{12}$ cm^{-3}; *2)* 10^{13} cm^{-3}.

In these experiments the dynamics of the current density distribution and of the electron heating over the discharge cross section was studied (during the current pulse). It was found that the current initially flows in a thin peripheral shell of the plasma, but this shell eventually expands toward the axis, and an essentially flat distribution is ultimately established. The distribution of the electron temperature over the discharge cross section behaves in a generally similar way (Fig. 32); i.e., neither a pronounced skin effect

FIGURE 32 Time evolution of the distribution of the electron temperature over the plasma column in the TTT device.[191]

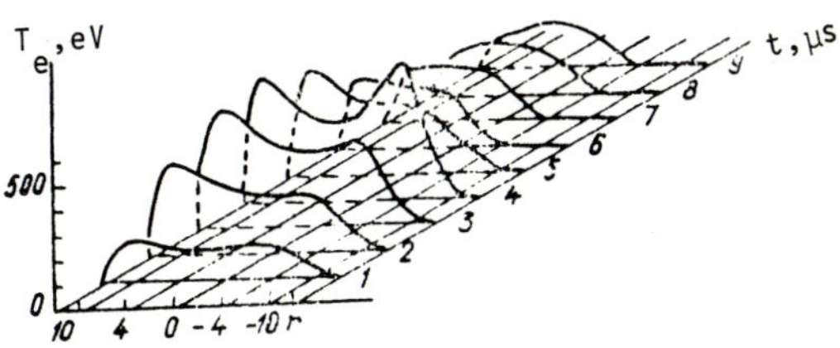

nor a concentration of the current and the temperature in peaks is observed. These results were obtained at a plasma density of 10^{13} cm^{-3}.

It should be noted that these experiments were actually carried out in a purely toroidal magnetic field, without a corresponding current in the plasma, which would have provided the necessary rotational transform after the turbulent-heating current pulse.

Experiments on the implementation of turbulent heating were carried out in 1974 in the TTF device in Frascati, Italy[250]. This is a toroidal magnetic confinement system of the tokamak type. A turbulent-heating electric field was produced around the loop by means of a sharp rise of the current in the plasma and reached a maximum value on the order of 25 V/cm. The current in the plasma reached its maximum (5 kA) in a few tens of microseconds and then decayed slowly to about 4 kA. The electron temperature determined from laser-scattering measurements was 5-20 eV, where the first of these figures corresponds to the preexisting plasma. At the same time, an x-ray procedure revealed the existence of a group of electrons with a temperature of 2.5-4 keV, having a clearly expressed anisotropy ($T_{e\parallel} > T_{e\perp}$). There was only a slight anomaly in the plasma resistance.

It follows from an analysis of the results that under those experimental conditions the current density and the electric field were not far enough above the critical values to generate a pronounced turbulence. By way of comparison we might cite the experiments of Ref. 69 (with similar tokamak parameters), where effective turbulent heating was achieved in fields six times the value of E_{max}

in the TTF, at current densities in the plasma nearly five times higher.

The anisotropy of the high-temperature group of electrons which was observed in these experiments suggests that the average energy of a directed stream of accelerated electrons was being detected.

A study of the turbulent heating of a rather dense plasma ($n = 7\cdot10^{13}$ cm^{-3}) in the T5-M tokamak[69] (1976) revealed some interesting aspects of the process when the system contains fast electrons. In this experiment a high-density "fast current" pulse with a strong loop electric field (150 V/cm) was superimposed on the slow current in the plasma which produced the magnetic field configuration required for confinement. During the flow of this fast current, a pronounced anomaly in the discharge resistance (up to 3 Ω) and a heating of the plasma were observed. Laser scattering revealed an electron temperature of 300 eV, and the plasma energy reached $4\cdot10^{16}$ eV/cm^3 according to measurements with a diamagnetic probe after the heating pulse. Measurements of the x-ray emission from limiters yielded estimates of the energy (about 250 keV) and the current (about 100 A) of the accelerated electrons in the system and also the role played by these electrons in the plasma energy balance. It was concluded from these estimates that the plasma heating could not be a result of a dissipation of the energy of the accelerated electrons.

Collective interactions in a current-carrying plasma have also been observed in experiments in plasma betatrons. As early as 1964 it was discovered that electrons could not be accelerated to relativistic energies in a dense plasma because of the intense plasma oscillations, which opposed the acceleration. These experiments[157] were carried out at the Khurchatov Institute in a toroidal betatron with strong loop electric fields (150-250 V/cm) and with a toroidal magnetic field. The current in the plasma became purely "ohmic." Energy measurements of the x-ray emission which occurred in these experiments revealed that the plasma electrons are heated to about 50 keV by the short electric-field pulse (with a length on the order of 10^{-7} s) under these conditions.

Collective interactions in strong electric fields are also observed in linear plasma betatrons,[133] but here the events are similar to those in beam–plasma discharges, which are not the subject of this discussion.

This brief review of the most important experiments on the anomalous plasma resistance and on turbulent heating has thus demonstrated that, despite the wide variety of sizes and types of magnetic confinement systems (mirror systems, uniform magnetic fields, stellarators, tokamaks, and "purely toroidal" fields), and despite the wide variety of plasma parameters, the events follow a common course in all cases: Small-scale rf instabilities are driven in a strong electric field in the plasma. These instabilities are responsible for the anomalous electrical conductivity and for the dissipation of directed-motion energy; i.e., they are responsible for the conversion of energy supplied by the electric field produced by the external source into thermal energy of the plasma electrons and ions.

CHAPTER III
KINETIC INSTABILITIES IN A CURRENT-CARRYING PLASMA

1 COLLECTIVE INTERACTIONS IN A PLASMA WITH A CURRENT

The extensive research on gas discharges has shown that collective interactions strongly affect the plasma properties in such systems. Even Langmuir, working many decades ago, was aware that collective oscillations are responsible for the randomization (Maxwellization) of electrons in a glow discharge at very low neutral-gas pressures.[246] Subsequent analysis confirmed that the path traced out by a charged particle in a plasma may be changed not only by binary collisions (Coulomb, ionizing, and other collisions of particles) but also in interactions of waves with the electric fields which arise in a nonequilibrium plasma. This effect is of practical importance because the plasmas in many experiments are in nonequilibrium states for one reason or another.

The governing role of waves in the interaction of an electron beam with a plasma was first demonstrated experimentally in some well-known studies by Merrill and Webb[255] and Looney and Brown.[249]

The simplest elementary mechanism for the collective interaction of an electron beam with a plasma is Landau damping. A particle moving along the propagation direction of a plasma wave, at a velocity near the phase velocity of the wave, exchanges energy with the wave rapidly. Whether it transfers energy to the wave or takes energy from it depends on the ratio of the wave's phase velocity to the particle's velocity. The resultant effect of the involvement of many particles naturally depends on how many more fast particles (which transfer energy to the wave) than slow particles (which damp the wave) there are in the electron velocity distribution. If the distribution is Maxwellian, the wave will be damped and the particles accelerated. If there is a directed stream of particles in the plasma, moving at a velocity slightly higher than the wave phase velocity, the wave will be intensified and the particles slowed. If a large number of waves are excited simultaneously in the plasma, and if the phases of these waves are coupled with each other only weakly, then the interaction of the particles with these waves will be a random one, having much in common with binary collisions. In this case we may introduce the concept of an effective collision rate (or frequency) and the concept of a collective electrical conductivity of the nonequilibrium plasma.

It follows that collective processes in a gas discharge may prove extremely important not only for reaching an understanding of the physics of plasmas but also for several applied problems, e.g., for heating current-carrying plasmas under conditions such that Coulomb interactions are unimportant. Collective interactions may also significantly affect other transport coefficients; in particular, they may lead to an elevated diffusion and thermal conductivity across a magnetic field. Collective interactions in plasmas absolutely must be taken into account in the development of plasma accelerators, and the effective excitation of intense electrostatic oscillations in a gas discharge may be exploited to develop high-power, extremely simple sources of millimeter and submillimeter waves. Collective interactions may play a major role in plasmas in space; for example, streaming instabilities may be primarily responsible for the radio emission of the sun, other stars, and the planets of the Solar System. A systematic experimental study of collective processes in nonequilibrium plasmas was begun around 1958 with a study of the beam–plasma interaction[39,77,174] which had been predicted and analyzed theoretically by Akhiezer and Faĭnberg[12,13] and by Bohm and Gross[198] in 1949.

2 *INTERACTION OF CHARGED-PARTICLE BEAMS WITH PLASMAS*

Since the primary purpose of this monograph is to analyze the experimental research on current-driven instabilities, and since an interaction of directed electron streams with plasmas underlies these instabilities, we will first consider the basic physics of the collective processes excited by an electron beam in a plasma. This is the simplest case of collective processes in a plasma, and it is of major interest not only for learning about the particular instability involved but also as an approach for modeling current-driven instabilities. This type of electron–electron interaction is the one which has received the most detailed study to date, both theoretically and experimentally.[165,166,248]

Beam–plasma interactions are usually observed in experiments involving the injection of electron beams into plasmas. The first such experiments were carried out in Refs. 39, 77, and 174. This instability may also be manifested in high-current gas discharges, either straight discharges, because an electron beam forms near the cathode[117,122,133,153] at a low neutral-gas pressure (10^{-3}-10^{-4} torr), or toroidal discharges, where a "beam" forms as the result of a continuous acceleration of electrons from the tail of a Maxwellian velocity distribution.[37,157]

The collective processes associated with beam—plasma instabilities are discussed in detail in the exhaustive reviews by Fainberg.[165,213] Here we will look at only certain aspects of this problem which will help explain the physical nature of the collective interactions in a plasma with a current. The most interesting case here is that of longitudinal current flow, with $\mathbf{v}_n \| \mathbf{H}_0$ ($\mathbf{v}_n$ is the directed velocity of the beam electrons). Furthermore, in research on collective interactions in very dense plasmas the condition $\omega_{ce} \ll \omega_{pe}$ essentially always holds. Accordingly, a magnetic field cannot have a governing effect on these processes.

Theoretical work on the excitation of longitudinal plasma waves by a charged-particle beam has shown that the instability is particularly pronounced if the directed velocity of the plasma electrons satisfies[13] $v_b > v_{Te}$. This interaction is based on the longitudinal-wave Cerenkov (or "Vavilov—Cerenkov") effect. If $n_1 < n$, where n_1 and n are the number densities of the beam electrons and the plasma electrons, respectively, the spectrum of rf waves which are excited is determined by the Cerenkov-excitation conditions and lies near ω_{pe}. In the hydrodynamic stage of the instability, while the energy spread in the beam is quite small, so that the condition $\frac{\Delta v}{v_n} \ll \left(\frac{n_1}{n}\right)^{1/3}$ holds, the growth rate of the wave amplitude can be written

$$\Gamma_0 \approx \sqrt{3} \cdot 2^{-4/3} \omega_{pe} \left(\frac{n_1}{n}\right)^{1/3}.$$

The solution of the problem in the linear approximation can thus be used to determine the conditions for an instability as well as the frequency and growth rate of the waves. With increasing wave energy, however, the waves begin to strongly influence the properties of the beam and the plasma, giving rise to an effective heating of the beam and plasma electrons, as has been shown experimentally. The spreading of the energy spectrum of the beam electrons causes a deviation from the condition for a resonant interaction of the electrons with the wave for a substantial fraction of the particles, and the efficiency of the beam–plasma interaction naturally falls off.

This inverse effect on the beam properties and the efficiency of the collective interaction can be studied in the quasilinear theory developed by Vedenov, Velikhov, and Sagdeev[55] and Drummond and Pines.[209] The collisionless relaxation of an initially monoenergetic beam in a plasma was studied by Shapiro,[179,178] who showed that a rather small fraction of the energy lost by the beam, proportional to $\left(\frac{n_1}{n}\right)^{1/3}$, goes into increasing the thermal energy of the beam during the initial, hydrodynamic, stage of the instability. Since the beam density is quite low, however, the beam temperature may deviate substantially from the plasma temperature. As a result the beam ceases to be "monoenergetic" at quite small wave amplitudes:

$$\sum_k |E_k|^2 \sim n_1 m_e v_n^2 \left(\frac{n_1}{n}\right)^{1/3}.$$

The spreading of beam velocities is pronounced: $\frac{\Delta v}{v_n} > \left(\frac{n_1}{n}\right)^{1/3}$. The growth rate of the plasma waves thus falls to

$$\Gamma_0^{(1)} \approx \omega_{pe} \frac{n_1}{n} \left(\frac{v_n}{\Delta v}\right)^2$$

in the quasilinear stage. With a further increase in the amplitude of the plasma waves, the distribution function of the beam electrons flattens out over an even broader velocity interval. The beam energy is converted into the electric-field energy of the wave and into the kinetic energy of the plasma particles. The interaction comes to a halt when the distribution function f_0 becomes a plateau, $\frac{\partial f_0}{\partial v}=0$, over the velocity interval from $v_1 = v_{Te}$ to $v_2 = v_n\left[1+\left(\frac{n_1}{n}\right)^{1/3}\right]$. The time over which the beam is slowed and the waves excited in the quasilinear stage is

$$t_{slow} \approx \frac{1}{\omega_{pe}} \frac{n}{n_1} \left(\frac{\Delta v}{v_0}\right)^2 \ln \frac{W_k}{W_0},$$

where $\Delta v = v_1 - v_2$, v_0 is the initial electron beam velocity, and W_k and W_0 are the final and initial wave energies. The distance over which the beam expends energy on exciting plasma waves can be estimated roughly to be

$$l_{slow} \approx \frac{v_0}{\Gamma_0} \approx 10^{-8} \frac{W}{j} \sqrt{n},$$

where W is the energy of the electron beam, in electron volts, and j is the beam current density, in amperes per square centimeter. The energy density in the plateau of the beam with the energy spread is given approximately by

$$W_e \approx \int_0^\infty f(v) \frac{m_e v^2}{2} dv \approx \frac{1}{3} W = \frac{1}{3} n_1 \frac{m_e v_n^2}{2}.$$

The energy lost by the beam as a result of the beam–plasma interaction is thus $W_{0e} \approx \frac{1}{3} n_1 \frac{m_e v_0^2}{2}$. It follows from quasilinear calculations by Shapiro and Shevchenko[177] that the energy lost by the beam is expended in equal parts on increasing the energy of the electrostatic waves,

$$\widetilde{W}_k = \frac{1}{8\pi} \sum_k |E_k|^2 \approx \frac{1}{3} n_1 \frac{m_e v_0^2}{2},$$

and increasing the kinetic energy of the plasma electrons.

A beam—plasma instability thus produces intense plasma waves in a discharge; these waves are converted by various mechanisms and give rise to an intense superthermal rf emission from the discharge and an effective heating of plasma electrons.

The imposition of a strong magnetic field parallel to the velocity of the beam electrons causes several new effects, e.g., the excitation of waves with frequencies near harmonics of the electron cyclotron frequency[156] and an intense heating of ions during a beam plasma instability. Even a weak magnetic field will shift the characteristic frequency of the beam—plasma instability toward the upper hybrid frequency.

Up to this point we have been considering an unbounded system. The finite radius of a system has the effect (on the one hand) of causing some decrease in the frequency of the plasma waves and some increase in the relaxation times and (on the other) of possibly driving a beam—drift instability in an inhomogeneous beam in a magnetic field, as was first shown by Mikhaĭcovskiĭ[138] and Lovetskiĭ and Rukhadze.[131] The mechanism underlying this instability is a convection of charges under the influence of the magnetic field and the perturbing electric field.

Waves can also be excited by a low-density beam if the velocity of the beam electrons is lower than the thermal velocity of the plasma electrons but higher than the propagation velocity of an ion acoustic wave, v_s, if the condition $T_e \gg T_i$ holds (if the plasma is "nonisothermal"). This instability, however, is of no practical importance in experiments on the injection of external beams into plasmas, since its growth rate is extremely small. It should be noted that this instability may determine the plasma properties essentially completely, in contrast with beam–plasma experiments in gas discharges.

We have been discussing the basic characteristics of the beam–plasma instability in the case in which the density of the beam electrons is much lower than the plasma density. As the beam density is raised, an instability occurs in the system once the critical density is passed (provided, of course, that the main condition, $v_n > v_{Te}$, holds). The plasma ions play an important role in this instability, in contrast with the preceding case, and this instability is called the electron–ion Budker–Buneman instability.[42,201] The critical beam current required for this instability to occur can be determined from the following expression under the condition $\omega_{ce} > \omega_{pe}$:

$$I_{cr} \equiv \pi a^2 n_1 e v_n = \frac{m a^2}{4e} \frac{\mathbf{k}^2 v_0^3}{\left[1 + \left(\frac{m_e}{M_i} \frac{\mathbf{k}^2}{k_z^2}\right)^{1/3}\right]^3}.$$

Here a is the beam radius, $\mathbf{k}$ is the total wave vector, and k_z is the projection of the wave vector onto the beam velocity direction. The driving of this instability in a beam–plasma interaction has been analyzed in detail in a review by Nezlin.[139]

The Budker–Buneman instability, it should be noted, plays an extremely important role in collective interactions in a gas discharge, but there both the conditions for its occurrence and its manifestations are slightly different. In a discharge the condition $n \approx n_1$ essentially always holds, so that the main criterion for the occurrence of the instability becomes the condition $v_n > v_{Te}$. Furthermore, in the course of the instability electrons are accelerated in the discharge electric field, and this acceleration can cause a deviation from resonance and significantly reduce the instability growth rate.

3 *INSTABILITY OF A CURRENT-CARRYING PLASMA IN A STRONG ELECTRIC FIELD*

Let us assume a nonisothermal plasma ($T_e >> T_i$) in a strong electric field in which the electron current drift velocity u is greater than the ion thermal velocity V_{Ti} but less than v_{Te}: $v_{Te} > u \geq v_s$. Under these conditions, as Gordeev[66] was the first to show, the induced Cerenkov effect can excite electrostatic ion acoustic waves with a frequency spectrum $\omega(k)$ ranging from the ion plasma frequency ω_{pi} to the reciprocal of the electron transit time in the system, $\frac{v_0}{L} < \omega(k) < \omega_{pi}$ (Refs. 121, 120, 147, and 66):

$$\omega(k) = kv \pm \omega_S, \quad \omega_S \approx \frac{kv_S}{\sqrt{1+k^2\rho_D^2}}.$$

This instability results from a resonant interaction of a wave with electrons which are moving at a velocity approximately equal to the phase velocity of the wave.

The growth rate for the ion acoustic instability is the

difference between the rate of the growth due to the electrons and the rate of the damping due to the ions:

$$\Gamma_S = \Gamma_{Se} - \Gamma_{Si} = \sqrt{\frac{\pi}{8}\frac{m_e}{M_i}}\frac{\mathbf{k}\mathbf{u} - \omega_S}{(1+k^2\rho_D^2)^{3/2}} \cdot$$

$$\exp -\left(\frac{\mathbf{k}\mathbf{u} - \omega_S}{\sqrt{2}k_{\parallel}v_{Te}}\right)^2 - \frac{\sqrt{\pi}\omega_S^2(\omega - \mathbf{k}\mathbf{v}_i)}{2k^2v_T^2}\exp - \left(\frac{\omega - \mathbf{k}\mathbf{v}_i}{\sqrt{2}k_{\parallel}v_{Ti}}\right)^2.$$

At $T_e \approx T_i$, and if $\mathbf{u} \parallel \mathbf{H}$, the ion acoustic instability cannot occur because of the damping by ions. Ion acoustic waves can thus be excited only in a very nonisothermal plasma with $T_e >> T_i$. Such a situation may arise, for example, in a weakly ionized gas in which ions lose energy at an appreciable rate in collisions with atoms of the cold neutral gas. This condition also holds frequently in time-varying discharges in which the ions do not manage to draw enough heat from the electrons to maintain the same temperature. Furthermore, a similar situation can arise because of the preferential heating of electrons in a current-carrying plasma during collective interactions or upon the injection of a plasma across a magnetic field.[101]

Gordeev[66] showed that the growth rate of ion acoustic waves under the condition $T_e >> T_i$ reaches a maximum at the frequency $\omega \approx 0.6\omega_{pi}$ given by

$$\Gamma_{S\,\max} \approx 0{,}38\sqrt{\frac{\pi}{8}\frac{m_e}{M_i}}\,\omega_{pi}\frac{u}{v_S}. \qquad \text{(III.1)}$$

The length of the waves, λ_s, is on the order of ρ_D.

It can be seen that the growth rate of the ion acoustic waves is significantly lower than the growth rates for other types of current-driven instabilities. In a gas discharge, however, if the electrodes have an adequate emission capability, the density of the electron stream approaches the plasma density. In this case the ion acoustic instability can be important if the rate of binary collisions of the plasma particles is sufficiently low ($\nu_{ei,\,em} \ll \Gamma_S$). Furthermore, this instability occurs at rather low electron current drift velocities, at which no other current-driven instability can compete.

The critical value of the electron current drift velocity for the excitation of ion acoustic waves depends strongly on the ratio T_e/T_i and the velocity distribution of the plasma particles. Figure 33 illustrates the stability with respect to the excitation of electrostatic ion acoustic waves for a current-carrying plasma which can be described by a double Maxwellian distribution. The numerical data used to plot this diagram were taken from Ref. 261. Rudakov and Korablev[147] have derived an analytic expression for the critical electron drift velocity for the ion acoustic instability for the case $T_e/T_i \gg 1$:

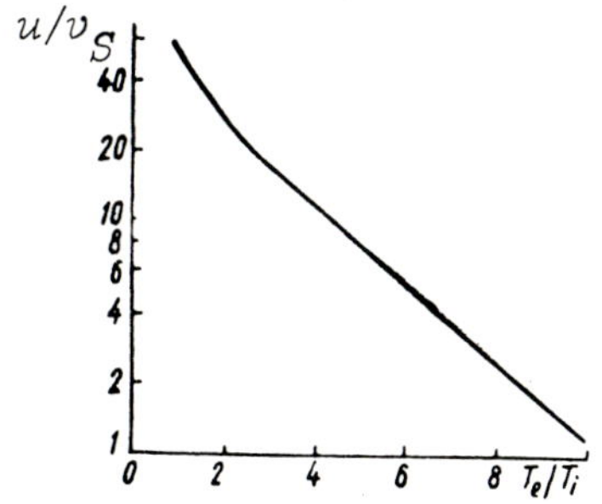

FIGURE 33 Critical drift velocity for driving the ion acoustic instability, expressed in units of the acoustic velocity, vs the ratio of the electron and ion temperatures.

$$u_{cr} \geqslant v_S \left[1 + \left(\frac{M_i}{m_e} \right)^{1/2} \left(\frac{T_e}{T_i} \right)^{3/2} \cdot \exp - \frac{1}{2} \left(\frac{T_e}{T_i} + 3 \right) \right] \equiv \alpha \left(\frac{T_e}{T_i} \right) v_S.$$

In practice, the condition $v_{cr} \approx v_s$ holds only at $T_e/T_i \geq 10$.

The critical current density required for the excitation of ion acoustic waves in the plasma can be written

$$j_{Scr} = nev_{Te}\sqrt{\frac{m_e}{2M_i}}\,\alpha\left(\frac{T_e}{T_i}\right).$$

The electron thermal velocity to be used in this expression is that in the last stage of the heating. It should be noted that the coefficient α is very sensitive to the energy distribution of the plasma particles.

As the current density in the plasma is raised, the conditions for the electron–ion Budker–Buneman hydrodynamic instability become satisfied as soon as the electron drift velocity becomes higher than the electron thermal velocity.[201] The critical current density for this instability may be written

$$j_{BB_{cr}} \geq nev_{Te} \gg j_{Scr}.$$

An important aspect of this instability is that there is no preferential ion damping, so that this instability can occur in an isothermal plasma, at $T_e \leq T_i$. This instability can take two forms.[280] At resonance, under the conditions $k_z u \approx \omega_{pe}$ and $\left|1-\frac{k_z u}{\omega_{pe}}\right| \leqslant \left(\frac{m_e}{M_i}\right)^{1/3}$, the growth rate reaches a maximum:

$$\Gamma_{BB\max} \approx \frac{\sqrt{3}}{2^{4/3}}\sqrt[3]{\frac{m_e}{M_i}}\,\omega_{pe}. \qquad \text{(III.2)}$$

For a hydrogen plasma, $\Gamma_{BB\max} = 0.0562\omega_{pe}$. The frequency of the electrostatic waves in this case is approximately equal to the Doppler-shifted electron plasma frequency of the electron beam. These waves lie in a very narrow wave-vector interval, while their frequency spectrum is rather broad:

$$\omega_{BB} \approx (1 \div 0.1)\,\omega_{pe}\sqrt[3]{\frac{m_e}{M_i}}. \qquad \text{(III.3)}$$

In a gas discharge the resonance $\omega_{pe} \approx k_z u$ is weakened by the electron acceleration in the discharge electric field.[280] The resonance condition is disrupted in a time of order $\Delta t \approx 1/\Gamma_{BBmax}$. At $|k_z u| < \omega_{pe}$, however, the nonresonant branch of the Budker–Buneman instability, with a growth rate much smaller than that in (III.2), can occur in the discharge:

$$\Gamma_{BB} \approx \frac{\omega_{pi}}{\sqrt{(\omega_{pi}/kv)^2 - 1}}.$$

This instability was first studied by Kovrizhnykh and Rukhadze.[121] Its frequency can be found only by taking a nonlinear approach to the problem. In order of magnitude, this frequency is near the ion plasma frequency, but the wavelength interval for this instability is very broad.

All these comments apply to a gas discharge in a magnetic field if $\mathbf{E} \parallel \mathbf{H}$ and if the magnetic field is weak enough to satisfy $\omega_{ce} \ll \omega_{pe}$. Conditions of this sort are satisfied well, for example, in the dense plasma of a high-current gas discharge.

4 COLLECTIVE INTERACTIONS AND CONDUCTIVITY OF THE PLASMA OF A GAS DISCHARGE

The electrical conductivity of a plasma is the basic factor determining the conversion of the energy of the directed electron drift in the electric field into the energy of random motion, i.e., the thermal energy of the plasma. It is one of the most important characteristics of the behavior of a gas-discharge plasma.

Spitzer and Harm[283] calculated the conductivity of a fully ionized gas under the assumption that the electric field in the plasma is weak enough that the energy acquired by an electron from the electric field over a mean free path is small in comparison with the electron thermal energy. As was mentioned in Chapter I, these calculations also incorporate the fact that electron–electron collisions are not capable of rapidly randomizing (Maxwellizing) the electrons, so that the electron velocity distribution in the presence of the electric field cannot be described by a shifted Maxwellian distribution.

The conductivity predicted for a fully ionized plasma by the expression based on binary interactions alone agrees well with experiments carried out in weak electric fields. However, as was first shown experimentally,[158] the plasma conductivity in strong electric fields is a complicated function of the electric field and the plasma density, and it depends weakly on the electron temperature. The reason for this behavior is that a variety of current-driven instabilities in the plasma result from collective interactions of the stream of electrons with the plasma. This fact was first demonstrated experimentally in Refs. 160 and 159, where a superthermal microwave emission (ω_{pi} to ω_{pe}) from a plasma was observed; this emission could be explained only on the basis of current-driven instabilities. The most important of the current-driven instabilities, which strongly affect the plasma conductivity, are the ion acoustic and electron—ion instabilities.

The theoretical results on the turbulent resistance of plasmas during various instabilities have been examined in detail in monographs by Vedenov and Ryutov,[57] Sagdeev and Galeev,[60] and Zavoĭskiĭ and Rudakov.[91] Here we will simply mention a few of these theoretical results which will be of assistance in interpreting the experiments described below.

As was mentioned earlier, in the nonisothermal plasma ($T_e > T_i$) of a gas discharge, ion acoustic waves are excited at a sufficiently high electron current drift velocity, $u > u_{cr}$. The resonant interaction of the electrons with an ion acoustic wave leads to a rapid increase in the amplitude of this wave, and in turn this amplitude increase leads to an inverse effect of the wave on the drift velocity. This effect takes the form of the appearance of a collective frictional force. As the electric field in the plasma is increased, the energy of the ion acoustic waves increases, and the collective frictional force begins to balance the effect of the electric field, reducing the electron drift velocity to a minimum value u_{cr} (Refs. 147 and 216).

As the electrons interact with the ion acoustic waves and undergo a quasilinear scattering, which tends to make the distribution function isotropic, the drift velocity thus automatically stabilizes at the level $u_{cr} = \alpha v_S$.

Galeev and Sagdeev[60] derived a simple relationship between the effective collision rate and the level of the turbulent electric-field fluctuations resulting from the current-driven instability, W_S: $\nu_{eff} = (\frac{W_S}{nT_e})\omega_{pe}$. With this expression and nonlinear estimates of the amplitudes of various instabilities, one can estimate the effective collision rate and thus the turbulent electrical conductivity of a plasma. The resulting expression can easily be tested experimentally, since all the quantities in it can be measured directly.

To derive an exact analytic solution for the electrical conductivity of a turbulent plasma is an extremely complicated and difficult task, since the kinetic equation for the waves and particles, incorporating the nonlinear scattering of waves by electrons and the scattering of electrons by turbulence fluctuations in collisions with ions and with each other, is a complicated integrodifferential equation. The kinetic equation derived for the waves from the weak-turbulence theory is applicable only in order of magnitude. Naturally, estimates of the turbulent noise level and the effective dynamic-friction force are also good only in order of magnitude. On the other hand, the simple and physically clear arguments and results of the quasilinear theory lead to quite accurate explanations for many experimental facts even without a detailed study of the nature of the turbulence. In several cases it is sufficient to make the obvious assumption that the turbulence is excited after a threshold value of the drift velocity is reached and that the inverse effect of the wave on the electron stream causes a large dynamic-friction force. In practice, this instability could be any one which draws its energy from the directed motion of the electrons. The turbulent conductivity can then be written

$$\sigma(E) = \frac{nev_S}{E} . \tag{III.4}$$

The complicated dependence of the turbulent conductivity on the various plasma parameters is actually hidden in the coefficient α and in the relationship between the electron temperature and the electric field in the plasma.

As mentioned earlier, the drift parameter α depends strongly on the velocity distribution of the plasma particles. For example, Vekshtein and Sagdeev[58] examined the distortion of the high-energy part of the ion energy distribution caused by the heating of some of the particles through the preferential Landau damping of ion acoustic waves, for the case of a collisionless shock wave. They found that the effect of the distortion is to increase the critical drift velocity by a factor of $\alpha = (4M_i/m_e)^{1/4}$ with respect to v_S: $u_{cr} = v_{Te}\sqrt[4]{\frac{m_e}{M_i}}$. For hydrogen, the critical velocity is

$u_{cr} = 9.3v_S$; i.e., $\alpha = 9.3$. This is the correct value of the critical drift velocity only when the plasma is undistorted by the energy loss from the confinement system. When such a loss occurs, u_{cr} is reduced, so that the plasma conductivity is reduced during the ion acoustic instability in the range of applicability of this approach.

The results derived for the conductivity agree well with measurements in straight high-current gas discharges at low electric fields and also at shock fronts.[96,106] In the latter example the electric field in the plasma is related to the plasma temperature by the condition for an equilibrium between the gas pressure and the magnetic pressure at the shock front; and the field is automatically held near the threshold for the instability, at the level $E_{cr}=\frac{neu_{cr}}{\sigma}$. Using $E=\frac{v_A}{c}H$ and $nT=\frac{H^2}{8\pi}$ for a shock wave, we find $(E)|E_{cr}=\frac{v_A}{c}\sqrt{8\pi nT_e}$. The plasma conductivity at the shock front is then given by

$$\sigma=\frac{\omega_{pe}}{4\pi}\frac{c}{v_A}\left(\frac{m_e}{M_i}\right)^{1/4}.$$

The values found for the coefficient α in various experiments on turbulent plasma heating lie in the range 1-20, depending on the discharge parameters (n, E, T_e, T_i). These results are consistent with the predictions generated by the theory for the ion acoustic instability under the assumption, for example, that some of the ions are heated.

In strong electric fields the dynamic-friction force may not be strong enough to balance the effect of the electric field. As a result, the current velocity will not stabilize at the level αv_s and will instead begin to increase slowly with increasing electric field. The simple estimates above are not valid in this case.

It has been shown experimentally that the plasma conductivity stops decreasing with increasing electric field and reaches a constant value which depends on only the nature of the gas and the plasma density.[76,20,152]

A plateau may be formed on the function $\sigma(E)$ by, for example, a current—convective instability, which occurs in strong electric fields and which greatly increases the energy flux out of the discharge along the magnetic field. In this case the function $\sigma(E)$ is[116]

$$\sigma_{\text{plateau}} \approx \frac{1}{4\pi}\left(\frac{M_i}{m_e}\right)^{1/2} \omega_{pe}.$$

The effect may be seen most clearly in straight discharges.

This discussion of the plasma conductivity has dealt with the case in which the electric field is not very weak, and the drift velocity of the electrons is lower than their thermal velocity. The conductivity of a plasma in which there is a strong electric field, such that the Budker–Buneman hydrodynamic electron–ion instability ($u > v_{Te}$) can occur, has been calculated in a series of papers.[152,179,178,201,281]

Shapiro[179] studied the anomalous electrical conductivity of a plasma caused by the scattering of electrons by the turbulent electric-field fluctuations of the Budker–Buneman instability. In this case the conductivity depends weakly on the electric field; for a hydrogen plasma it can described by $\sigma_{BB}^{(1)} \approx 31\omega_{pe}$.

Hamberger and Friedman[222] determined the conductivity in the case of the Budker–Buneman instability by working from the assumption that the effective collision rate is on the order of the growth rate for the resonant branch of the instability: $\sigma_{BB}^{(2)} \approx \frac{1}{2} \cdot \left(\frac{M_i}{m_e}\right)^{1/3} \omega_{pe}$. For a hydrogen plasma, $\sigma_{BB}^{(2)} = 6.15\omega_{pe}$.

Sizonenko and Stepanov[280] showed that the resonant branch of the Budker–Buneman instability is weakened in a sufficiently strong electric field in a gas discharge, since the resonance is disrupted by the acceleration of electrons in the discharge electric field in a time considerably shorter than the rise time of this instability. In this case an anomalous conductivity may result from the scattering of electrons by turbulent electric-field fluctuations corresponding to the nonresonant instability branch.

The resonant and nonresonant branches of the Budker–Buneman instability lead to approximately equal numerical results for the conductivity of a hydrogen plasma. The dependence of the conductivity on the mass of the working gas, however, is quite different in the two cases: $\sigma_{BB} \sim \left(\frac{M_i}{m_e}\right)^{1/3}$ for the resonant branch and $\sigma_{BB} \sim \left(\frac{M_i}{m_e}\right)^{1/2}$ for the nonresonant branch. Since the turbulence is strong in this case, it is a very complicated matter to derive the exact characteristics of the turbulent state for such a plasma; the difficulties have yet to be overcome. The results which have been derived are thus reliable only in order of magnitude.

It should also be noted that a plasma in a highly turbulent state may prove unstable with respect to high-frequency, purely electron waves with frequencies and growth rates on the order of ω_{pe} and with a characteristic wave vector $k \approx \frac{\omega_{pe}}{u}$. This instability can occur when a multistream motion is set up in an electron gas, and it may also be driven by a nonuniformity of the electron velocity profile in a highly turbulent state. This comment may help explain why in experiments on turbulent plasmas one observes not only low-frequency waves, at frequencies ω_S and ω_{BB}, but also high-frequency waves at frequencies near ω_{pe}. These waves cannot be explained by the linear theory of electrostatic current-driven instabilities.

Another possible explanation is the formation of a low-density electron beam, either near the cathode of a discharge with electrodes or through the acceleration of electrons from the tail of a Maxwellian distribution in a toroidal electrodeless discharge.

Various effects which can be explained only on the basis of the Budker–Buneman instability in the plasma of a high-current gas discharge have been observed experimentally by many investigators (see Refs. 76, 152, 160, and 222, for example). In particular, Suprunenko et al.[289] have shown that in a high-current gas discharge this instability acts to trigger the excitation of ion sound in an originally isothermal plasma. A more detailed experimental study of this instability has been carried out by Hamberger[222] in the plasma of a pulsed high-current toroidal discharge.

CHAPTER IV
ANOMALOUS CONDUCTIVITY AND TURBULENT HEATING IN STRAIGHT SYSTEMS

1 *STATEMENT OF THE PROBLEM*

The anomalous electrical resistance of a plasma was first observed experimentally in straight discharges with electrodes. This would be quite natural to expect, since in toroidal systems, in which the loss is determined by the thermal conductivity across the magnetic field, the increase in the effective collision rate which accompanies collective interactions not only increases the amount of energy which is pumped into the plasma but also increases the energy loss, while in straight systems with a strong longitudinal magnetic field, where the loss is determined primarily by the longitudinal thermal conductivity, the increased collision rate reduces the energy loss.

In Chapters I-III we saw that the anomalous electrical conductivity is a very complex phenomenon, a manifestation of effects which differ widely in physical nature but which are related by the common nature of the collective interactions in plasmas.

One of the many difficulties complicating efforts to compare the various experiments with the theory is that the course of events depends on the external conditions of the discharge and on the properties of the particular magnetic confinement systems, especially the method used to produce the initial plasma, the electric and magnetic fields, and the behavior of the energy loss. What generally happens in a given experiment is a succession of several different but related processes which appear superficially to occur as a single effect. The nature of the anomalous resistance may also be affected strongly by MHD instabilities, which establish the geometric configuration of the plasma column and its density profile.[300] In turn, the anomalous conductivity affects the nature of MHD instabilities in a discharge (resistive modes).[217,240] This interrelationship between microinstabilities and macroinstabilities causes yet a further complication of the external picture of the phenomenon. The parameters of the discharge circuit and the power provided by the power supply also determine the conditions under which this phenomenon occurs. A switch from an approximately constant current to an approximately constant voltage may substantially change the nature of the collective interactions in the plasma.[195]

At present, we do not have any single theory which is capable of describing the overall phenomenon. In an analysis of experiments, the phenomenon must therefore be examined in connection with all the processes and characteristics of the particular experiment. Although the ion acoustic instability has a small growth rate, it may occur at low values of the electron drift velocity, at which other instabilities do not occur; by limiting the drift velocity, the ion acoustic instability may prevent the attainment of higher thresholds for other instabilities. It is accordingly the ion acoustic instability which determines the nature of the turbulent conductivity of the plasmas in many experiments. [In this case it may be assumed that the drift parameter α in Eq. (III.4) is nearly constant.]

It is usually a simple matter to produce a nonisothermal plasma as the starting point for an experiment, either by making use of the time variation of a discharge (in which the ion temperature cannot rise fast enough to keep up with the electron temperature) or by keeping the ionization of the working gas at a weak level (so that ions readily lose energy in collisions with neutrals). The difference between the ion and electron temperatures may be increased even further by the preferential heating of electrons in a plasma with a current.

Direct measurements in several experiments have confirmed that the current drift velocity is indeed stabilized during the ion acoustic instability (Fig. 34; see Refs. 45, 79, and 290, for example).

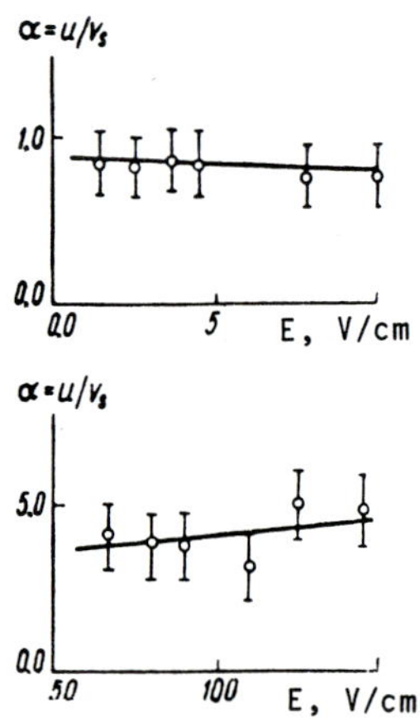

FIGURE 34 Dependence of the drift parameter on the electric field in the plasma, demonstrating the stabilization of the drift velocity by the ion acoustic instability.[51]

A comparison of the experimental values of the electrical conductivity of plasmas with theoretical predictions requires much caution, since the expression for the conductivity requires the instantaneous values of various parameters, which are usually very difficult to measure. For a comparison of the final values of the conductivity, which are customarily measured, it becomes necessary to consider the relationship between the electron temperature and the electric field – a relationship determined by the confinement properties of the particular magnetic confinement system.

We will now consider several examples and compare the experiments which have been carried out with a simple phenomenological theory.

2 *TURBULENT HEATING IN A STRAIGHT DISCHARGE*

To examine the turbulent heating of a plasma in a straight discharge, we take the approach of a phenomenological analysis. We will analyze how the electrical conductivity behaves under the conditions prevailing in a high-current straight gas discharge stabilized by a longitudinal magnetic field.

To take the effect of collective interactions on the conductivity into account we introduce only a single assumption: that the current drift velocity may be stabilized at the phase velocity of the wave excited by the current. As we have seen, this assumption is supported well by experiment. The drift velocities observed experimentally have usually been quite low, so that the waves which have been excited could only have been ion acoustic waves.

The experimental $\sigma(E)$ dependence can be found explicitly only by considering the particular form of the energy loss from the discharge on the electrodes. This problem can be solved for a straight discharge which is stabilized by a longitudinal magnetic field in the case of a weak electric field in the plasma, so that it is legitimate to ignore the anomalous loss due to the current–convective instability, with a growth rate proportional to E (Ref. 108).

We first assume that the mean free path of the electrons between collisions with each other, λ_{ee}, is considerably shorter than the distance between the electrodes, l, and we assume that the electron Larmor radius is considerably shorter than the radius of the plasma column. In this case the energy loss may be determined from the expression for the classical electron thermal conductivity along the magnetic field, by using the turbulent collision rate.

Taking Braginskii's approach, we write the complete energy balance equation in a straight plasma column in a strong longitudinal magnetic field as follows (strictly speaking, this equation holds only for the axial region)[41]:

$$nT_e \operatorname{div} u + \frac{3}{2}\frac{dnT_e}{dt} = -\operatorname{div} q + jE - 0.71nu\frac{\partial T_e}{\partial z}. \qquad \text{(IV.1)}$$

This equation reflects the changes caused in the plasma density and pressure in the discharge by both the heat fluxes and the thermal emf. For a high-current gas discharge the expression for the heat fluxes must incorporate not only the diffusion flux but also a loss due to heat removal by the stream of electrons:

$$q = -\varkappa_{\parallel}^{e} \operatorname{grad} nT_e + 0.71nT_e u.$$

For simplicity we restrict the discussion to

steady-state solutions, with $\frac{\partial nT_e}{\partial t} = 0$. In this case the continuity equation

$$\frac{\partial n}{\partial t} + \mathrm{div}\, un = 0 \tag{IV.2}$$

leads to the relation nu = const, which means that the current density remains constant along the length of the system. Furthermore, in order to determine the electric field in the plasma we must use the equation of motion[41]

$$m_e n \frac{du}{dt} = -\frac{\partial nT_e}{\partial z} - enE - 0.71 n \frac{\partial T_e}{\partial z} + \frac{j_\parallel}{\sigma_\parallel} en. \tag{IV.3}$$

Using (IV.2) we can simplify (IV.3) to a form accurate within the assumptions $\frac{\partial \ln n}{\partial z} = -\frac{1}{2}\frac{\partial \ln T}{\partial z}$ and $u \ll v_{Te}$:

$$1.21 \frac{\partial T}{\partial z} + eE - eE_0 = 0,$$

where $E_0 = \frac{V_0}{l}$.

The electron thermal conductivity along the magnetic field can be written in the form[41]

$$\varkappa_\parallel^e = 3.16 \frac{T_e}{m_e \nu_{ee}}. \tag{IV.4}$$

We substitute the turbulent collision rate ν_{eff} into this expression:

$$\nu_{eff} = \frac{\omega_{pe}^2}{4\pi\sigma_{turb}}. \tag{IV.5}$$

Let us consider the case of ion acoustic turbulence, in which the current drift velocity is stabilized near the ion acoustic velocity:

$$nu = an\sqrt{\frac{T_e}{M_i}}. \qquad \text{(IV.6)}$$

Using continuity equation (IV.2) and (III.4), we can write the energy balance equation and the equation of motion in forms convenient for computer calculations:

$$p^2 \frac{d}{dx} \frac{\Theta(x)}{\varepsilon(x)} \frac{d\Theta(x)}{dx} + 2.71p \frac{d\Theta(x)}{dx} + \varepsilon(x) = 0, \qquad \text{(IV.7)}$$

$$\varepsilon(x) = 0.96p \frac{d\Theta(x)}{dx} - 1. \qquad \text{(IV.8)}$$

Here $x = \frac{z}{l}$; $\Theta(x) = \frac{T_e(x)}{T_0}$; T_0 is the maximum electron temperature at the discharge axis; $\varepsilon(x) = \frac{E(x)}{E_0}$; and $p = 1.26T_0/eE_0l$. Knowing p, we can relate the electron temperature in the central part of the discharge to the potential difference between the electrodes for the case of turbulent plasma heating in a straight gas discharge, and this result can easily be tested experimentally: $T_0 = p\frac{V_0}{1.26}$.

For an exact solution of system (IV.6)-(IV.8) we must specify the boundary conditions at the boundaries of the plasma column. In practice, these conditions are known only poorly because of the complexity of the events which occur in the layer at the plasma–electrode boundary. At the very least we should expect large drops in both the plasma density and the plasma temperature here. In solving the system of equations, however, we can calculate the parameter p for arbitrary boundary values of the temperature and determine the relationship between p and the end temperatures. It turns out that a variation of the boundary conditions over the reasonable range from $\Theta_0 = 0$ to

$\Theta_0 = 0.5$ changes the parameter p from 0.33 to 0.5, i.e., by a factor of less than two. This accuracy in the value of the electron temperature in the central part of the discharge is quite adequate for a comparison of the calculations with experiments. Figures 35 and 36 show the functions $\Theta(x)$ and $\Theta(x)$ found through a numerical solution of the system of equations for the value $\Theta_0 = 0.5$ at the cathode end of the plasma column. The plasma temperature reaches its maximum far over on the anode side of the discharge; this result sounds reasonable in view of the intense stream of electrons from the cathode toward the anode, caused by the strong electric field. The temperature drops so rapidly near the anode that the parameter p becomes essentially independent

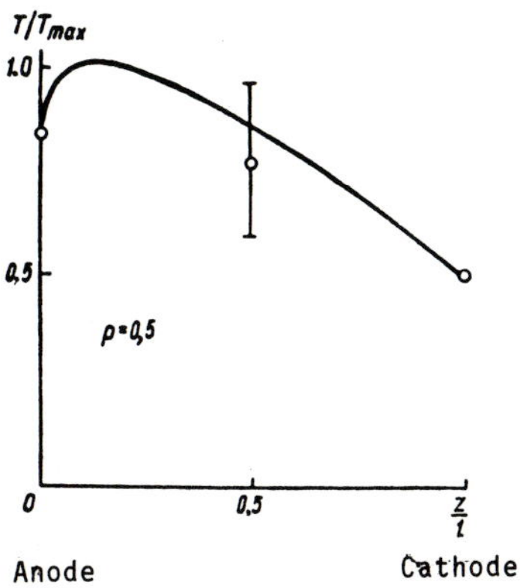

FIGURE 35 Electron temperature profile along the axis of a straight discharge with electrodes during turbulent plasma heating in the GROM-3. The curve was found through a numerical solution of Eq. (IV.7).

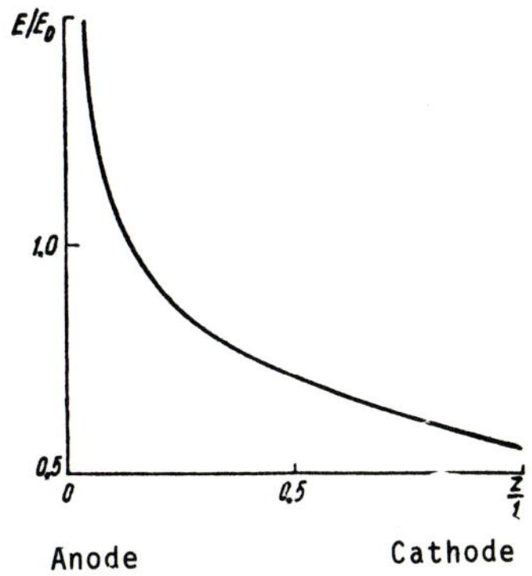

FIGURE 36 Electric field profile in a straight discharge with electrodes according to a numerical solution of Eq. (IV.8).

of the anode temperature (a boundary value). Near the cathode, in contrast, the temperature varies quite smoothly.

The electric field can be assumed to remain approximately constant over a large distance along the discharge, equal to $E_{cr} \approx 0.6E_0$. This uniformity results from the thermal emf which arises when there is a longitudinal temperature gradient. Very near the anode the electric field rises in absolute value to $10E_0$, causing a significant decrease in the potential in the anode part of the plasma column: $\Delta V_a \approx 0.4V_0$. Figure 37 shows the longitudinal profile of the potential at the axis of the column, measured with an electrostatic probe in the high-current gas discharge in the GROM-3 device under turbulent-heating conditions.[289] The potential drop near the cathode which is observed experimentally is a consequence of processes which occur at the cathode and has been ignored in these calculations. Figure 35 shows the experimental values of the electron temperature obtained with an x-ray probe in the GROM-3. The good agreement between experiment and calculations here lends support to the assumptions embodied in the calculations and implies that the thermal emf plays an important role in the evolution of a straight high-current turbulent discharge.

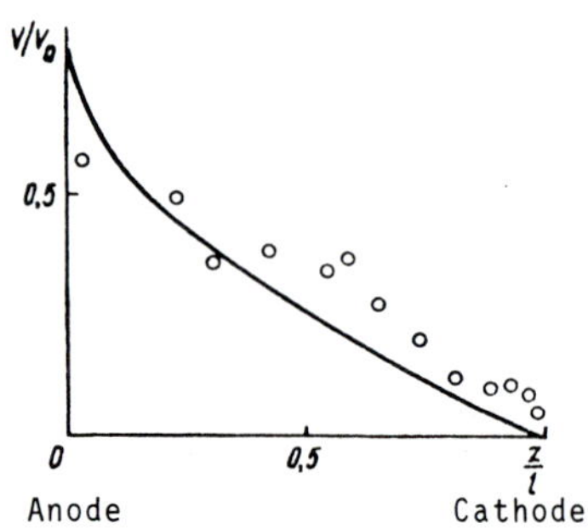

FIGURE 37 Axial potential profile in a turbulent straight gas discharge according to a solution of Eq. (IV.8). The circles show experimental data from the GROM-3 device.

3 *PLASMA CONDUCTIVITY IN A STRAIGHT GAS DISCHARGE*

The relationship between the maximum temperature in the central region of a high-current gas discharge and the potential difference between the electrodes was found quite accurately in Section (IV.2):

$$T_0 = \frac{eE_0 l}{2.52} . \tag{IV.9}$$

A corresponding expression can be derived for the case of a Spitzer conductivity. At first glance, Eq. (IV.9) might appear to imply that the final plasma temperature does not depend on whether the conductivity is anomalous but only on the potential difference between the electrodes of the straight discharge. The turbulent-heating effects can be seen only during the briefer heating interval, calculated from the time-varying solution of Eq. (IV.7). In practice, however, the discharge current is set by the external circuit rather than by the electric field in the plasma. In this case it is easy to show that with I_{pl} = const we have

$$\frac{T_{turb}}{T_{coll}} = \frac{\sigma_{coll}}{\sigma_{turb}} = \frac{1}{\alpha}\left(\frac{2M_i}{m_e}\right)^{1/2}\frac{E}{E_{Dr}};$$

i.e., for a given current in the plasma, a much higher temperature is reached in the turbulent discharge if the electric field in the plasma exceeds a certain value

$$E_{cr} = E_{Dr}\alpha\left(\frac{m_e}{2M_i}\right)^{1/2}$$

(for hydrogen, $E_{cr} \approx 10^{-14}(n/T_e)$. Here E_{Dr} is the Dreicer field: that at which all the plasma electrons should have been caught up in continuous acceleration if there had been no collective effects.[208] With $E = E_{Dr}$ and $\alpha = 1$, for example, we would have $T_{turb} = 60T_{coll}$ in a hydrogen plasma.

If the electric field in the plasma is set by the external circuit, the turbulent heating may be very slight, despite a highly anomalous conductivity. The relationship between the electron temperature and the discharge current under turbulent conditions can easily be found from the condition for excitation of ion sound, (III.1):

$$I_{cr} \geqslant \alpha n e \sqrt{\frac{T_e}{M_i}}\, \gamma^* \pi a^2 \qquad \text{(IV.10)}$$

where the coefficient γ^* represents the effective cross-sectional area of the current, which is related to the radial profile of the current: $S_{eff} = \gamma^* \pi \alpha^2$. Expression (IV.10) can easily be transformed to

$$\beta_{\varphi} = \frac{nT}{H_{\varphi}^2/8\pi} = \frac{7.9 \cdot 10^{24}}{\omega_{pe}^2 \; a^2} \qquad \text{(IV.11)}$$

Despite the variety of conditions under which turbulent plasma heating is studied in pulsed straight discharges, analysis of the experimental results yields values of β_{φ} which lie in a narror interval, [0.2, 1.0] (see Fig. 38, which is taken from Ref. 83). It follows that β_{φ} is a characteristic parameter of the particular experiment. The highest value, $\beta_{\varphi} = 1$, was obtained in experiments with a high plasma density, and the lowest value, $\beta_{\varphi} = 0.2$, was obtained with a low density.

Substituting the expression for the electron temperature in a turbulent straight discharge, (IV.9), into the expression for the conductivity of a turbulent plasma, (IV.6), we find the function $\sigma(E)$ for the particular type of energy loss:

$$\sigma_{turb}(E) = \frac{an}{\sqrt{E}}\sqrt{\frac{le^3}{M_i}} = 1.1 \cdot 10^{-13}\frac{n\sqrt{l}}{\sqrt{E}}. \qquad \text{(IV.12)}$$

It is important to note that $\sigma \sim E^{-1/2}$. This is the $\sigma(E)$ dependence which was observed in the GROM-1 experiments,[158] which were described in detail in Chapter II. Figure 39 shows the dependence of the plasma conductivity in the high-current gas discharge in the GROM-1 on the square root of the electric field in the plasma, measured with probes. We see that the functional dependence $\sigma(E)$ predicted by the calculations agrees well with the experimental results.

As a convenient measure of just how anomalous the plasma conductivity is, we adopt the ratio $\sigma_{coll}/\sigma_{turb}$, which may be written for a straight discharge as

$$\frac{\sigma_{coll}}{\sigma_{turb}} = \frac{0.42 \cdot 10^{14} E^2 l}{n}.$$

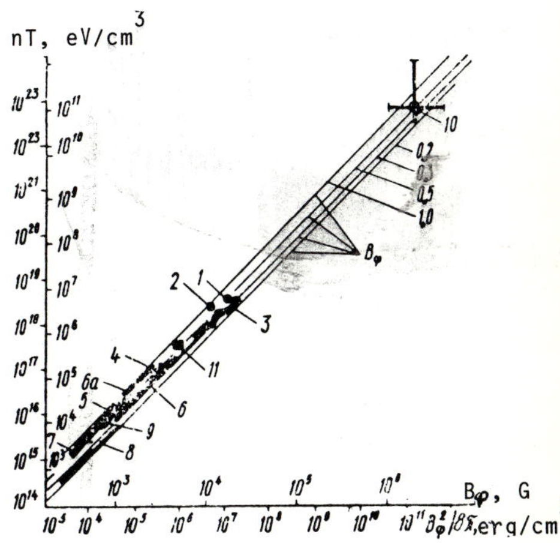

FIGURE 38 Dependence of the plasma pressure on the magnetic field produced by the current in various experiments. 1,2) Experiments with a stabilized pinch[123,200]; 3,4,5,6,6*a*,7) experiments on turbulent heating in straight discharges[135,259,114,83,239]; 8,9) experiments in the T-3 and TM-3 tokamaks (Refs. 38 and 218); 10) experiments in plasma-focus systems.[170]

The fact that relations (IV.9) and (IV.12) are used for ordinary experimental conditions, $\sigma_{coll} \approx 14uT_e^{3/2}$, is taken into account here.

Ordinarily, all the measurements are taken before the (m = 1) kink instability of the plasma column has had time to occur.

The distance between the electrodes in the GROM-1 device was l = 50 cm; the diameter of the plasma column was $2a$ = 8 cm; and the plasma density was $n \approx 3.5 \cdot 10^{14}$ cm^{-3}. The electric field was varied from 0.3 to 10 V/cm and thus remained well below the Dreicer field at all times. In this case the conductivity anomaly can reach a value $\sigma_{coll}/\sigma_{turb}$ = 600.

As the electron temperature changes from 15 to 150 eV, the critical Dreicer field changes from 200 to 20 V/cm. In this case we can expect the turbulent-heating effect to be quite weak $(T_{turb}/T_{coll}$ = 1-3). The basic purpose of this experiment was thus to study the behavior of the anomalous conductivity as a function of the electric field.

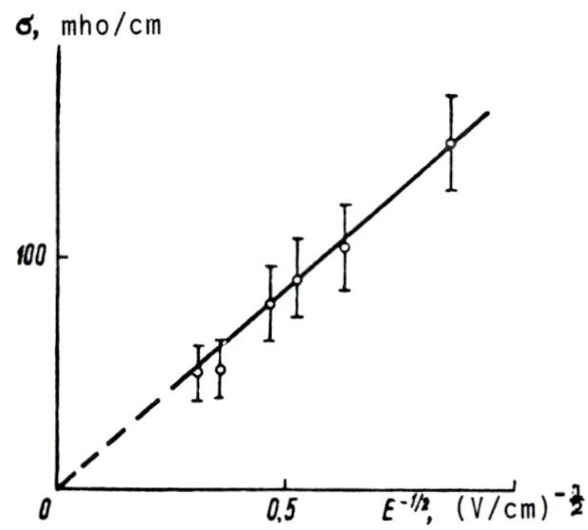

FIGURE 39 Conductivity of the turbulent plasma vs the square root of the electric field (in the GROM-1).

Figure 34 shows the experimental results on the drift parameter $\sigma = u/v_S$ (the ratio of the drift velocity to the ion acoustic velocity) for various electric fields in the plasma. Here the electron temperature is determined from the diamagnetic signal. We see that this ratio is approximately one and does not depend on the electric field, again in good agreement with the assumption of an ion acoustic instability.

The mean free path of the electrons between collisions with each other in these experiments lies in the range 2-100 cm for Coulomb interactions, while if collective interactions are taken into account it lies in the interval 0.2-0.03 cm. Substitution of the experimental parameters into Eq. (IV.12) yields $\sigma_{turb} = 270/\sqrt{E}$. Since the experimental results yield $\sigma = 180/\sqrt{E}$ (Fig. 39) it may be assumed that the two results are in satisfactory agreement.

It is easy to see that expression (IV.12) determines the maximum conductivity of the plasma in a straight discharge. Since the electron temperature was varied over the interval $[T_0/2, T_0]$, the coefficient in this expression should lie in the interval 203-270. Furthermore, all the calculations in this section of the book have been carried out under the assumption that the plasma density remains constant over time, so that when the plasma pressure reaches its maximum the plasma density is at its initial value. When the dynamics of the turbulent heating in the discharge is taken into account (further on in this chapter) the maximum temperature in the discharge decreases, so that the coefficient in (IV.12) decreases.

The assumption that the drift velocity stabilizes in the turbulent plasma of a straight discharge — an assumption which leads to the electric-field dependence of the conductivity in (III.4) — is therefore consistent with the experimental results, at least at $E < E_{Dr}$. It turns out that for an unmagnetized plasma we have $\alpha = 1$ at $\omega_{ce}/\omega_{pe} \approx 0.05$.

Because of the weak electric fields in the GROM-1, the experimental conditions were quite unusual. Because of the incomplete ionization of the working gas, the relationship $T_e >> T_i$ was established immediately, and the conditions corresponded to the inequality $u > v_s$. In other words, the conditions for the occurrence of the ion acoustic instability were arranged in their pure form. Nevertheless, these numerical calculations confirm the occurrence of the ion acoustic instability only in a phenomenological sense and therefore cannot be freely extended to other conditions. Experiments must be continued under these particular conditions for a more detailed study of the basic behavior and a comparison of this behavior with the phenomenological calculations. By "basic behavior" here we mean the dependence of σ, nT, and $\frac{u}{v_s}$ on the plasma density, the mass of the working gas, the magnetic field, etc. The spectrum and amplitude of the turbulent fluctuations in the discharge must also be studied.

In concluding this section we wish to point out that the electrical conductivity of a plasma is determined by the rate of change of the momentum during electron–ion collisions, ν_{ei}, while its thermal conductivity is determined by the energy exchange during electron–electron collisions, ν_{ee}. These two rates are numerically equal in the case of binary Coulomb conditions, but they are not always equal for a turbulent plasma, since interactions occur through waves. These calculations have also used the assumption that the effective turbulent collisions are similar to Coulomb collisions and are equally effective in dissipating the momentum and arranging energy exchange between electrons. Unfortunately, the weak dependence $\sigma \sim \left(\frac{\nu_{ee}}{\nu_{ei}} \right)^{\frac{1}{4}}$ in these experiments ruled out a final conclusion regarding the effect of the ratio ν_{ee}/ν_{ei}.

4 ELECTRICAL CONDUCTIVITY OF THE PLASMA OF A STRAIGHT DISCHARGE AT STRONG ELECTRIC FIELDS

We have been discussing the electrical conductivity of a plasma at weak electric fields, in which case the electron thermal conductivity along the magnetic field can be described by (IV.4). Most experiments on turbulent plasma heating, in contrast, have been carried out at considerably higher electric fields, at which the energy loss from the discharge is determined by more-complex mechanisms. The reason for using strong fields is that only then can we expect the pronounced heating effect (Section 3) which is of interest in connection with controlled thermonuclear reactions. Furthermore, it turns out that when the electric fields in the plasma are strong the electrical conductivity of the plasma in a straight discharge becomes quite different. As the field is increased, the conductivity of the plasma in a system with a longitudinal energy loss initially falls off, but when a critical field, near the Dreicer field, is reached the conductivity reaches a plateau (Fig. 8).

Table III lists the basic results and the experimental conditions of experiments carried out in various laboratories around the world, not only in straight systems but also in closed systems without a toroidal equilibrium of the plasma. In this case, even in the closed systems, the energy loss results primarily from the thermal conductivity along the magnetic field. In this sense the closed systems are similar to open systems in terms of the characteristics with which we are primarily concerned here.

The first question which arises is this: Is the plateau on the function $\sigma(E)$ a consequence of a nonlinear saturation of the collective dynamic-friction force at strong electric fields, or is it instead a consequence of the particular nature of the energy loss from a straight discharge? In this connection it is important to examine the $\alpha(E)$ dependence (Fig. 34). Although the value of α does vary slightly from one experiment to another, its dependence on the electric field is always so weak that it could not explain the $\sigma(E)$ plateau. Apparently, then, the appearance of the plateau does not mean a pronounced nonlinear saturation of the turbulent dynamic-friction force, and the stabilization of the drift velocity at u_{cr} continues.

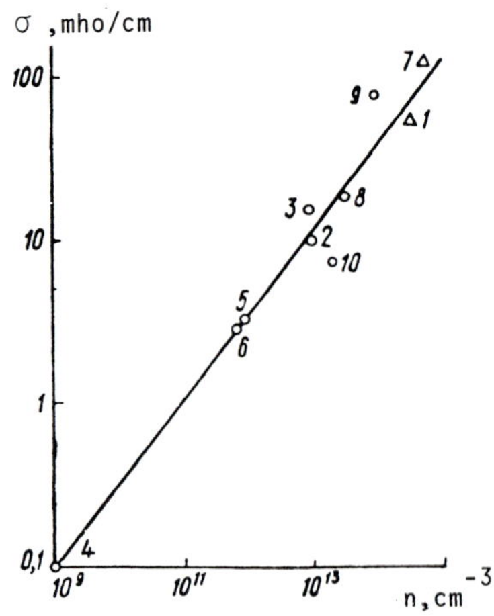

FIGURE 40 Electrical conductivity on the plateau vs the plasma density. The experimental points are labeled with the corresponding row in Table 3.

Working from the density dependence of the plasma conductivity shown in Fig. 40, we may say, with some reservation, that the expression

$$\sigma \approx 4.6 \cdot 10^{-6}\sqrt{n} \tag{IV.13}$$

holds over a rather broad range of parameters. The random excursions of the coefficients in (IV.13) from the average value,

Table 3

Summary of the results obtained in research on the turbulent heating and the anomalous resistance of current-carrying plasmas

No.	σ_{turb}, mho/cm	n, cm^{-3}	$\frac{\sigma}{\sqrt{n}} \cdot 10^5$	E_0, V/cm	nT_0/E_0^2, cm$^{-1}\cdot V^{-1}$	nT_0, eV/cm	Type of Discharge	Year	Ref.
1	50	$3.5\cdot10^{14}$	2.8	30	$4\cdot10^{13}$	$4\cdot10^{16}$	GROM-1 straight discharge	1961	158
2	10	10^{13}	3.2	--	--	--	Mirror	1965	153
3	12	10^{13}	4.6	--	--	--	SIRIUS stellarator	1966	43
4	7	$1.7\cdot10^{13}$	3.8	200	10^{12}	$4\cdot10^{16}$	Discharge in the **NPR-2 mirror**	1970	115
5	3.3	10^{12}	3.3	200	$2.5\cdot10^{11}$	10^{16}	VIKhR'-2 bumpy torus	1968	214
6	3.2	10^{12} — $4\cdot10^{13}$	3.3	500	$8\cdot10^{11}$	$2\cdot10^{17}$	TWIST stellarator	1968	222
7	120	$7\cdot10^{14}$	4.6	150	$8\cdot10^{13}$	$2\cdot10^{18}$	GROM-3 straight discharge	1969	152,135
8	18	$3\cdot10^{13}$	3.3	200	$5\cdot10^{11}$	$2\cdot10^{16}$	Toroidal discharge	1969	23
9	70	10^{14}	7	--	--	--	YaNUS-2 skin-layer straight discharge	1970	82
10	0.9	$2\cdot10^{12}$	0.6	200	$3\cdot10^{11}$	$2\cdot10^{16}$	Straight discharge in mirror	1969	33
11	0.6	$4\cdot10^{13}$	0.1	300	$2\cdot10^{12}$	$2\cdot10^{17}$	TURHE hollow-cathode straight discharge	1973	266
12	--	$7\cdot10^{14}$	--	300	$3\cdot10^{13}$	$3\cdot10^{18}$	GROM-4 mirror	1971	136
13	0.3	10^{14}	0.03	220	$.4\cdot10^{13}$	$7\cdot10^{17}$	THE MACH II mirror	1976	257
14	2.87	$5\cdot10^{13}$	0.4	83	$7\cdot10^{11}$	$5\cdot10^{15}$	BSG mirror	1976	257

in both directions and by as much as a factor of two, presumably result from a dependence of the plasma conductivity on other discharge parameters, which have not been taken into account here. One possibility is a dependence on the magnetic field, which was not monitored in these experiments.

Furthermore, in all experiments with straight discharges the dependence of the thermal energy density on the electric field in the plateau has been found to be quadratic:

$$nT_0 \simeq BE^2. \qquad \text{(IV.14)}$$

Here B is an empirical coefficient. Figure 41 shows the dependence $nT_0(E)$ found in experiments on turbulent plasma heating in the GROM-4 mirror system. Figure 42 shows values

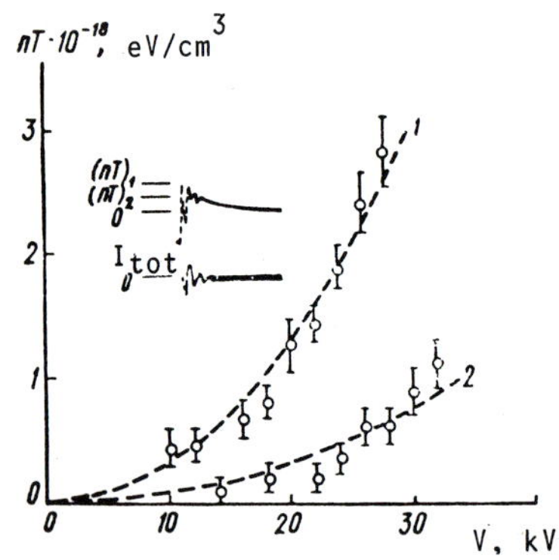

FIGURE 41 Plasma pressure vs the voltage to which the capacitor bank is charged in the turbulent heating of a dense plasma in the GROM-4 mirror. *1)* During the flow of the turbulent-heating current; *2)* after the current flow.

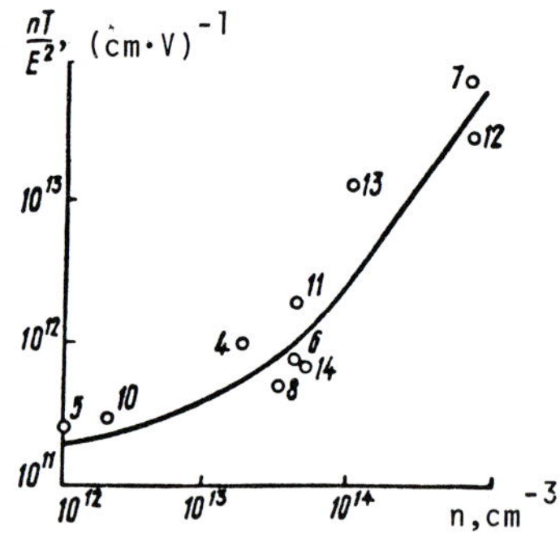

FIGURE 42 Dependence of the coefficient in Eq. (IV.14) on the plasma density found from an analysis of the results of various turbulent-heating experiments. The experimental points are labeled with the row number in Table 3.

found for the coefficient in (IV.14) under some very different experimental conditions. We see that this coefficient increases rapidly with increasing plasma density, so that it may confidently be asserted that over a broader range of parameters the dependence observed experimentally is described more accurately by $T_0 \propto E^2$ than by (IV.14). The efficiency of the turbulent heating increases with increasing plasma density.

We thus find that the three experimental facts embodied in (IV.14), (IV.14), and (III.4) are noncontradictory and are mutually complementary. Substituting (IV.14) into (III.4), we find (IV.14). This system of mutually consistent experimental facts can be used to construct a phenomenological description of the turbulent conductivity on the $\sigma(E)$ plateau at strong electric fields. In determining the experimental conditions it is important to note that all the experiments discussed in this section were carried out at a preheating electric field stronger than or equal to the critical Dreicer field. Nevertheless, the system was not observed to go into the regime of continuous electron acceleration, so that the relative number of runaway electrons remained at a level less than 1%.

Working from the experimental dependence in (IV.14), we can find an expression for the thermal conductivity from the steady-state solution of the energy balance equation: $\varkappa_{\parallel}^{e} \propto \frac{1}{\nu_{\text{eff}}}$. The inverse proportionality of the thermal conductivity and the effective collision rate which emerges from these experiments suggests that energy loss along the magnetic field is playing an important role [cf. (IV.4)].

Analysis[289,114] of the ways in which energy leaks out

of a hot plasma has shown that most of the loss (up to 70-80% of the energy provided by the power supply) occurs along the magnetic field to the anode. The same conclusion follows from the analytic calculation in Section 3.

Under the constraints imposed by the empirical relations in (IV.13), (IV.14), and (III.4), it is difficult to suggest a reasonable dependence of the coefficient in (IV.14) on the other discharge parameters which would put the thermal conductivity in its classical form, (IV.4), for a turbulent plasma in the plateau region. It is probably necessary to assume that an anomalous longitudinal thermal conductivity is acting here.

As Rudakov has shown,[116] the current-convective instability, which arises in a plasma with an anomalously high collision rate, can cause an anomalously rapid increase in the thermal conductivity of the plasma along the magnetic field and can also cause a removal of energy to the anode by hot electrons. This instability occurs in a plasma with a finite electrical conductivity in which a current is flowing; its growth rate is

$$\Gamma_{m,k}=-i\overset{e}{\varkappa}_{\parallel}\mathbf{k}^2+i\frac{k_yE_0c}{k_zH\partial j/\partial E}\,\frac{d\sigma_{\text{turb}}}{dr}. \tag{IV.15}$$

This expression reflects the possibility that the plasma conductivity may depend on the electric field. The first term in (IV.15) may be ignored because it is small in comparison with the second. Setting

$$\frac{d\sigma_{\text{turb}}}{dr}=\frac{\sigma_{\text{turb}}}{a},\quad \frac{\partial j}{\partial E}=\sigma_{\text{turb}}),\ k_y=\frac{1}{a},\quad k_z=\frac{\pi}{l},$$

we find

$$|\Gamma_{m,k}|=10^8\frac{V_0}{\pi a^2H}.$$

For voltages across the electrodes in the range 20-50 kV and for magnetic fields in the range 2-20 kOe we find $\Gamma^{-1} = 0.2\text{-}0.04\ \mu s$. In most experiments, therefore, this effect may occur.

The current—convective instability becomes particularly influential in a plasma containing a high energy:

$$nT \gtrsim \left(\frac{M_i}{m_e}\right)^2 \frac{E^2}{8\pi} \qquad \text{(IV.16)}$$

Condition (IV.16) actually determines a maximum amount of energy which a plasma may contain during turbulent heating before the current—convective instability becomes important. For hydrogen, expression (IV.16) may be written $nT \simeq 10^{12}E^2$. It should be noted here that (IV.16) is an estimate, and the results which it yields, while having the correct functional dependence, are valid only within an order of magnitude. When the current—convective instability occurs the current density may be redistributed over the cross section of the plasma column, and a region with a strong electric field may arise. This instability may also give rise to a stream of runaway electrons. Measurements have shown, however, that the current density in this stream is usually low in most experiments, no more than 0.1 of the total discharge current, and the energy carried to the anode by the stream over the entire course of the discharge does not exceed 10-20% of the total energy dissipated at the anode. Furthermore, after the onset of the current—convective instability of a turbulent plasma, intense microwave noise begins to be emitted near the plasma frequency and its second harmonic. It turns out that this noise should not occur in the case of the pure ion acoustic instability, and the explanation probably lies in the formation of a virtual cathode in this case.[117,133]

An important confirmation of Rudakov's theory might come from an experimental study of the dependence of the electrical conductivity of a plasma on the mass of the working gas [see (IV.16)]. As a rule, a dependence of this sort has not been observed over the mass range from hydrogen (1) to argon (140) (Ref. 84). One possible explanation is that the experiments were carried out in strong magnetic fields, in which the elementary mechanism for the turbulent resistance may be of a different physical nature.

5 DYNAMICS OF TURBULENT PLASMA HEATING IN A STRAIGHT DISCHARGE

Up to this point we have been discussing a hypothetical steady state of a straight turbulent discharge. It might be asked whether such a steady state is possible and, if so, how we are to reach it. In experiments in straight systems the signals from the diamagnetic and x-ray probes are pulsed signals, functions of the kinetic energy density in the plasma. They are usually electrical pulses with a typical rise time on the order of a fraction of a microsecond and a decay time on the order of several microseconds. They remain pulsed signals even if an approximately steady external electric field is applied to the plasma or if an

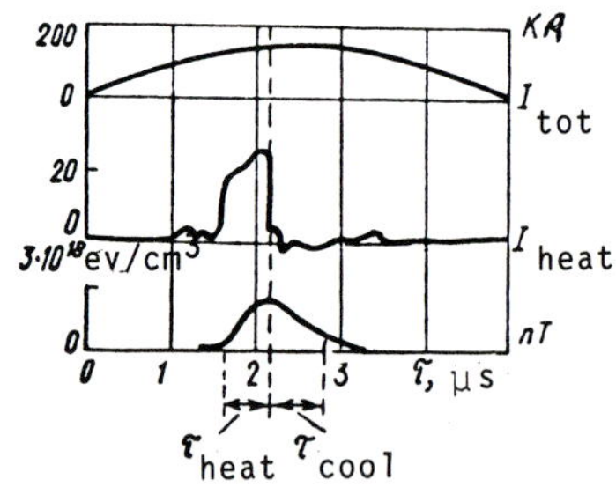

FIGURE 43 Oscilloscope traces of the total discharge current, the current at the axis of the plasma column, and the diamagnetic signal from experiments in the GROM-3 device.

approximately steady current pulse is passed through it. Figure 43 shows some representative oscilloscope traces of an approximately steady current pulse in a plasma and of the diamagnetic signal during turbulent heating. These observations, made at an early stage in the research, in fact made it possible to pose the following question: Is there the possibility in principle of a steady-state anomalous electrical conductivity in a current-carrying plasma? Experiments carried out in toroidal devices with good plasma confinement have furnished an absolutely unambiguous answer: Yes!

In experiments on turbulent plasma heating in straight systems, the rate of the ambipolar diffusion of particles toward the ends of the discharge typically increases as the electrons are heated. When the turbulent heating rate is much higher than the rate at which cool plasma is supplied, this effect causes the plasma density in the discharge to decrease over time.

The first factor in the product nT thus decreases rapidly, at a rate which increases with increasing electron temperature, while the second factor reaches a steady state after a rapid rise; the product, therefore, initially increases, reaches a maximum, and then begins to decrease, at a rate set by ambipolar diffusion to the ends. We turn now to some numerical estimates of this effect.

Let us consider the time-varying solution of Eq. (IV.1) at the point corresponding to the maximum temperature, where $\Theta_0 = 1$ and $\dfrac{\partial\Theta}{\partial t} = 0$. After some straightforward manipulations of (IV.1) we find

$$\frac{\partial\Theta}{\partial\tau} = \Theta^{1/2}(1-\Theta^2). \qquad \text{(IV.17)}$$

Here $\tau = t/\tau_0$, where $\tau_0 = pl/2.52v_S = 0.33 \cdot 10^{-6} l/\sqrt{V_0}$. The analytic solution of this equation is

$$\tau = \frac{1}{2}\log\left(\frac{1+\theta^{1/2}}{1-\theta^{1/2}}\right) + \text{arctg}\,\theta^{1/2}$$

and is shown in Figure 44. We now consider the equation for ambipolar diffusion of the plasma along the magnetic field[3]:

$$\frac{\partial n}{\partial t} = -\text{div}\, q_n \qquad \text{(IV.18)}$$

where $q_n \approx (v_S/\nu_{ei})\,\text{grad}\, n$. After the turbulent collision rate from (IV.5) is substituted into (IV.18), for the point corresponding to the maximum plasma temperature in the discharge, we find $\delta n/\delta\tau \approx -n$, where $\tau = t/6.35\tau_0$. The solution of this equation is $n = n_{0}e^{-t/635\tau_0}$.

Figure 45 is a plot of the product $\dfrac{n(\tau)T(\tau)}{n_0T_0}$, where n_0 is the initial plasma density in the discharge, and

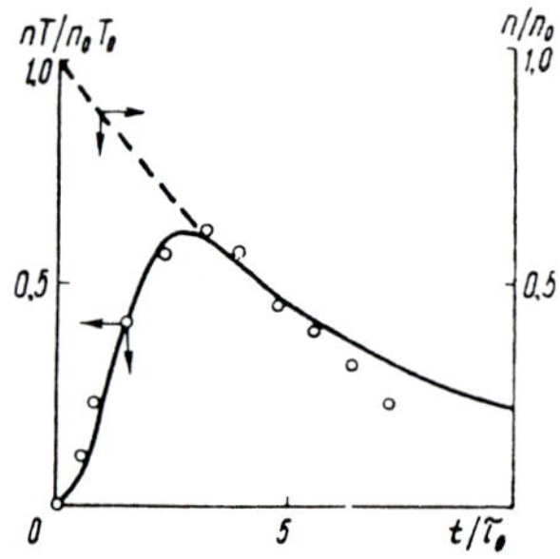

FIGURE 45 Time dependence of the plasma pressure demonstrating the dynamics of the turbulent heating in a straight discharge. The circles are data points obtained in direct measurements in Ref.33. $\tau_0 = 0.25\mu s$.

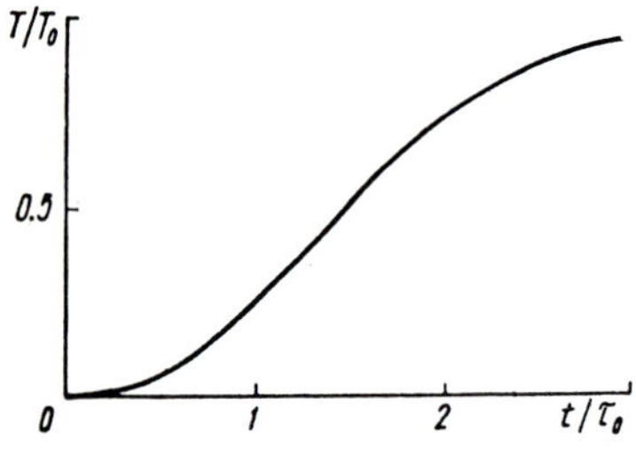

FIGURE 44 Temperature rise in a turbulent straight gas discharge according to a solution of Eq. (IV.17).

T_0 is the maximum electron temperature in the steady state. It is clear from Fig. 45 that at the time the kinetic energy in the discharge reaches its maximum (this energy is usually measured in experiments) the electron temperature has not yet reached its steady-state value, while the plasma density has dropped by a factor of nearly two.

It follows directly from these estimates that a steady state cannot be achieved in a straight turbulent discharge with end losses, since the maximum kinetic pressure of this type of discharge is inherently a pulsed effect.

6 TURBULENT HEATING OF IONS

This phenomenological discussion of turbulent heating has furnished some results which turn out to agree well with experiment. In this approximation, if only the binary interactions are taken into account the heating of the ions should result from a transfer of energy from turbulently heated electrons to ions at a rate

$$Q_{e\to i}=3/2\,\frac{T_e-T_i}{\tau_{eq}}\,n\approx 2{,}64\cdot 10^{-8}\,\frac{n^2}{AT^{1/2}}\quad [\mathrm{eV/(cm^3\cdot s)}]$$

at $T_e/T_i \gg 1.5$ (Ref. 9). For the range of plasma parameters typical of turbulent-heating experiments ($T_e = 3\cdot 10^3$ eV, $n \approx 10^{12}$-10^{15} cm^{-3}) the scale time for energy transfer from electrons to ions lies in the range 2.5-2.5$\cdot 10^{-3}$ s, while the turbulent-heating pulse is shorter by a factor of at least 10^3. From the standpoint of classical binary interactions, we would thus not expect any significant ion heating under these experimental conditions. Nevertheless, it was shown as early as 1966, in experiments in the PR-1 and

GROM-3 devices,[19,163] that turbulent heating by a longitudinal current is accompanied by an effective heating of the bulk of the ions to a temperature comparable to the electron temperature. The ion heating occurs under the same conditions and in the same time interval as the electron heating, and it is accompanied by the onset of an anomalously low electrical conductivity of the plasma and by intense superthermal rf emission from the discharge in the frequency range from ω_{pe} to a frequency below ω_{pi}. In some more-accurate experiments carried out later[221,245] it became possible to measure the delay between the application of the turbulent-heating current pulse and the appearance of the hot ions; it turned out to be at least 0.5 μs at a current pulse length of 5 μs. In this case the delay between the appearance of the hot electrons and the appearance of the anomalous conductivity, on the one hand, and the beginning of the current pulse, on the other, did not exceed 0.1 μs.

Table 4 shows the parameters of the most important experiments on turbulent ion heating. It is clear that these experiments can be classified in two groups.

In the first group there is an intense heating of both plasma components, so that a ratio $T_e/T_i<2$ is reached. The maximum ion temperature is probably set by a degradation of the thermal conductivity across the magnetic field, when the Larmor radius of the heated ions becomes comparable to the radius of the plasma column.

In the second, and rather small, group of experiments there is a preferential heating of the ions, so that $T_i >> T_e$. Characteristically, the experiments of the second group were carried out in a straight discharge with a grounded metal discharge chamber.

Table 4

Summary of results on turbulent ion heating in high-current gas discharges

T_i, keV	H_{min}, kOe	ρ_{Hi}, cm	a, cm	n, cm^{-3}	T_e, keV	T_e/T_i	a/ρ_{Hi}	Year	Ref	Comment
2.0	10-12	0.64	1.0	$7\cdot10^{14}$	3.6	1.8	1.5	1967	163	GROM-3
								1970	289	
0.5	1.6	2	6.0	$2\cdot10^{12}$	--	--	3	1968	239	Metal discharge chamber
50/700*	0.5-3.0	1	2.0	10^{14}	0.1	2.0	2	1970	82	
1.2-1.6	3.0	2	2.0	10^{12}	2-3	1.8	1	1967	221	TWIST
4.2	7.0	0.41	10	$1.4\cdot10^{13}$	--	--	25	1968	20	Metal discharge chamber
2.4	3.0	2.3	6-8	10^{13}	1.3	1.1	3	1967	90	TN-4
4	3.0	3	6	$10^{10}-10^{12}$	10	2.5	2	1965	153	
0.44/1.4*	17	0.17	7	$2\cdot10^{14}$	10	23	42	1976	243	OCU II
0.85/1.4*	17	0.24	7	$5\cdot10^{14}$	10	12	29	1976	243	OCU II
0.500	1.5	2.1	3	$2\cdot10^{11}$	1.0	2	1	1973	257 223	Working gas: helium
0.075/0.23*	12	0.1	5	10^{13}	0.7	19	50	1976	30	" " "

*These values correspond to the high-energy part of the ion distribution.

We first consider the experimental results from the first group, since they are the most typical of ohmic heating by longitudinal currents under conditions with an effective turbulent collision rate.

Ion heating occurs only when the electric field in the plasma exceeds a certain critical field which is approximately equal to E_{Dr} (Fig. 9). This point was demonstrated most clearly in the GROM-3 experiments.[289] They showed that the angular distribution of the bulk of the hot ions is generally

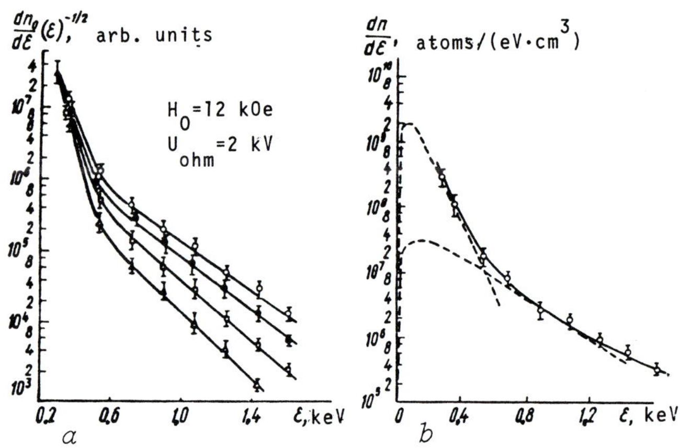

FIGURE 46 *a)* Dependence of the product $(dn_0/d\varepsilon)/\varepsilon^{1/2}$ on the energy (ε) of the charge-exchange neutral atoms at various times during turbulent plasma heating in the SIRIUS stellarator [o: 50 μs after the beginning of the discharge; ●: 125 μs; □: 200 μs; △: 300 μs after the beginning of the discharge]. *b)* Dependence of $dn/d\varepsilon$ on ε (solid curve) and reconstructed distribution functions demonstrating the two-temperature ion distribution function during turbulent plasma heating. The temperature of the cool component is 75 eV; that of the hot component is 230 eV; H_0 = 12 kOe; U_{ohm} = 2 kV; t_d = 50 μs.

isotropic, and the ion temperature reaches a maximum near the anode, so that the ratio T_e/T_i remains nearly constant over the length of the tube.

The ion energy distribution turned out to consist of two components (Fig. 46): the bulk of the ions, having a temperature approximately equal to that of the electrons, and a small group with a significantly higher temperature.[30,82,257] With increasing plasma density at a constant heating current, the temperature of the larger group of ions decreases, so that the product nT_i remains approximately constant; the effective temperature of the high-energy group of ions (the tail of the distribution function), on the other hand, remains essentially constant.[6] In the course of the heating, as the temperature of the bulk of the ions increases over time, there is an increase in only the density of the high-energy ions, not their energy.

As the longitudinal magnetic field is intensified, however, the energy of the bulk of the ions does not change, while that of the tail of the distribution function increases. Figure 47, from Ref. 186, shows that the energy of this tail increases nearly in proportion to the ion mass, while the number of ions in the tail decreases. The energy of the bulk of the ions is essentially independent of the ion mass. An important aspect of these experiments was the presence of ions with a different mass in the discharge; small numbers of ions of the mass of interest were added to the predominant ion species (helium).

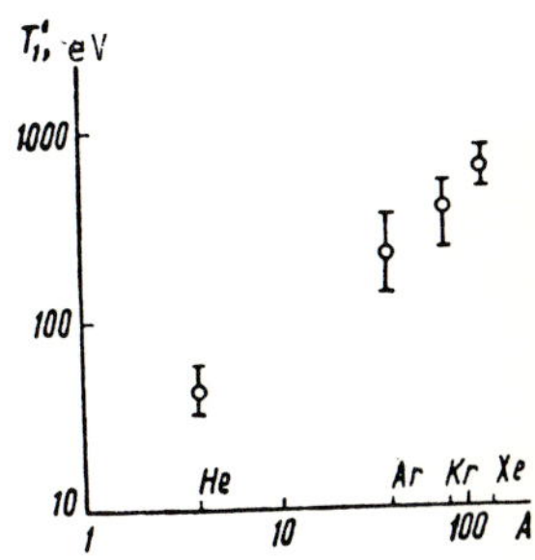

FIGURE 47 Dependence of the temperature of the hot component of the turbulently heated ions on their mass.[186]

These results suggest that there is probably some mechanism which accelerates the ions of the various masses to the same velocity. It was shown in the experiments of Ref. 30 that this velocity is approximately equal to the phase velocity of an ion acoustic wave (Fig. 48), regardless of the plasma density and the magnetic field. This behavior might confirm the suggestion that high-energy ions form as a result of acceleration in the field of an ion acoustic wave. The slow increase in the temperature of the tail ions with v_S, however, may indicate a different mechanism.

Study of the dependence of the energy of the bulk of the ions on the potential difference between the electrodes of a straight discharge (the most detailed study was carried out in Ref. 266) has shown that the ion temperature, like the electron temperature, is proportional to V_0^2, so that the measured ratio $n(T_e+T_i)/nT_i$ is approximately 4.5 and independent of the electric field in the plasma. The energy pumped into the plasma ions during the turbulent heating is thus proportional to I^2/σ . It should be recalled that under conditions such that ion heating is observed $(E > E_{Dr})$ in straight discharges the plasma conductivity ceases to depend on the electric field in the plasma, so that the function $\sigma(E)$ has a plateau.

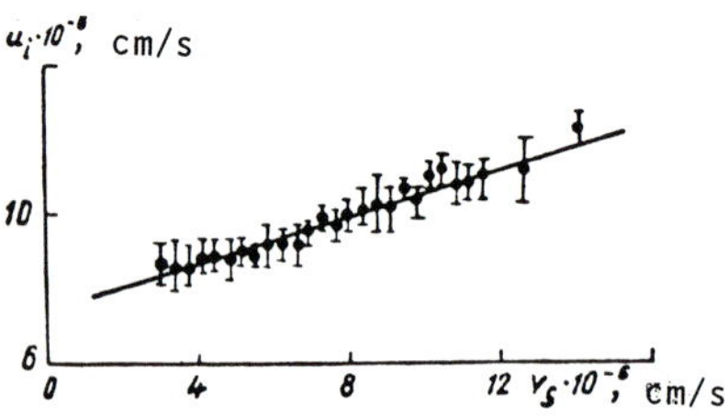

FIGURE 48 Dependence of the velocity of the hot component of the turbulently heated ions on the ion acoustic velocity from experiments from the SIRIUS stellarator.

It was mentioned earlier that the simple assumption that the current drift velocity is stabilized by current-driven instabilities in the plasma leads to a quantitative explanation for nearly all the microscopic phenomena associated with turbulent electron heating for a given value of the drift parameter. The only point which remains unexplained is the particular value of α which is reached for some turbulence mechanism or another. In this approach, ion heating is ignored. At the same time, the fact that there is an effective turbulent heating of ions, to temperatures comparable to the electron temperature, may now be regarded as proved beyond any doubt. Since the final ion temperature in this case is far higher than the initial temperature, it should be assumed that the heating is determined by very nonlinear effects in the discharge, so that a simple classical examination of the problem will naturally prove inadequate for reaching an understanding.

Calculations of the rate of change of the ratio T_i/T_e from the quasilinear theory of turbulent heating developed by Rudakov[147,91] have shown that at $T_e > T_i$ we have

$$\frac{T_i}{T_e} \approx \left[2\ln \left(\frac{2T_e}{T_i} \quad \frac{v_{T_e}}{v_{T_i}} \right) \right]^{-1}$$

For a hydrogen plasma, $T_i/T_e < 0.05$.

In terms of quasilinear effects, and if collisions are ignored, resonant ions can absorb only a small fraction of the energy supplied to the plasma. When the heating of the nonresonant ions (the bulk of the ions) is taken into account, however, we can expect this effect to be intensified substantially. During turbulent plasma heating by a current, therefore, the increase in the thermal

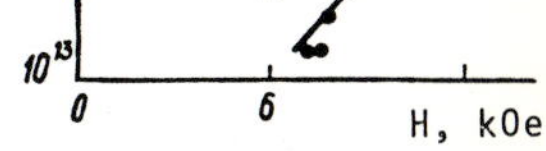

FIGURE 49 Magnetic field dependence of the total energy of the high-energy hydrogen ions in the URAGAN-2 stellarator. $nT_e \simeq 10^{15}$ eV/cm^3; $E = 0.1$ V/cm; $n = 6 \cdot 10^{12}$ cm^{-3}; $I = 6$–8 kA.

energy density of the bulk of the ions may be accompanied by the appearance of a small number of hot ions (a high-energy tail). The quasilinear theory furnishes a qualitative explanation for nearly all the basic phenomena associated with turbulent ion heating, but the numerical values predicted for α and T_i/T_e by these calaculations are much smaller than the experimental values.

Analysis of the onset of ion acoustic turbulence under the assumption that a broadening of the resonant region of the ion—wave interaction plays a major role also fails to explain the high experimental values of T_i/T_e (Ref. 91).

The ion heating rate calculated from the mechanism of nonlinear Landau damping turns out to be small.[91] A more important mechanism may be the trapping of ions by a wave excited in the plasma by the current. Numerical calculations[194,233] have shown that there may be an effective heating of ions by means of ion acoustic turbulence to temperatures near the electron temperature, but in several cases this mechanism also predicts values of T_i/T_e smaller than the experimental value. Furthermore, numerical calculations show that the ion heating should occur in a time of order $2\cdot 10^3/\omega_{pe}$, i.e., in the range 10^{-9}-10^{-7} s for the experiments which have been carried out, but the time observed experimentally frequently turns out to be much longer.[245] The linear dependence of the tail temperature on the mass of the admixed ions is also inconsistent with the resonance mechanism of acceleration by Landau damping.

In summary, the experimental results on turbulent ion heating cannot at present be explained completely on the basis of ion acoustic turbulence. It is probably reasonable to suggest that the effective ion heating in these experiments results from the driving of some other instability in parallel, possibly an instability associated with the current. The ion acoustic instability is stopped when the temperature of the turbulently heated ions approaches the electron temperature.

In most of the experiments with $T_i < T_e$ the turbulent ion heating can be explained on the basis of low-frequency waves excited in the current-carrying plasma. In the THE MACH II device, for example, double electrostatic probes near the axis of the straight turbulent discharge revealed low-frequency waves with a very high amplitude (frequency of 2-20 MHz, amplitude up to 100 V). These waves appeared suddenly and persisted throughout the ion-heating time interval.[243] Their nature was not affected by the orientation of the probe (along or across the magnetic field). In this case there may have been a stochastic heating of the ions.[31,287] Under the assumption that these low-frequency waves are approximately monochromatic with a frequency ω satisfying $\omega\tau^{*}>1$, where τ^{*} is the autocorrelation time of the waves, we can write the rate at which an ion acquires energy from the field of this wave as follows:

$$\frac{dT_i}{dt} = \frac{ze^2}{M_i} \cdot \frac{\langle E^2\rangle\tau^{*}}{(\omega^2\tau_{*}^2 + 1)}$$

Here $\langle E^2\rangle$ is the mean square field of the low-frequency turbulent fluctuations.

The high-energy ion component may result from (for example) a macroscopic motion of the plasma column as the result of a current—convective or MHD instability of the current column.[298] The linear dependence of the tail temperature on the gas mass agrees satisfactorily with the assumption that the macroscopic motion plays an important role.

As mentioned earlier, the experiments in which effective ion heating was observed, and in which values $T_i >> T_e$ were achieved, were carried out in grounded metal chambers. It is perfectly obvious that the ion acoustic instability could not occur in this case. The most probable mechanism here is the onset of instabilities in a rotating plasma in crossed electric and magnetic fields.[144] It would be completely natural to expect a radial electric field to arise in a straight discharge with a grounded metal chamber because of a potential difference between the plasma column and the walls. This mechanism has been studied in detail in a series of experiments in which a radial electric field was deliberately produced. The results show that this field is highly effective in heating ions.[105,187]

The turbulent ion heating in strong magnetic fields, with $\omega_{ce}/\omega_{pe} > 1$, in the TM-3 (Ref. 38) and ALCATOR[276] tokamaks with $E << E_{Dr}$ and in the URAGAN-2 stellarator[79,132] at $E > E_{Dr}$ has been linked to the excitation in a current-carrying plasma of a time-varying superthermal noise spectrum in the frequency interval near the lower hybrid resonance, $\omega \sim \omega_L \approx \omega_{pi}$. A likely heating mechanism is a modulational instability in this frequency range.* In all cases the drift velocity stabilizes at a level below the electron thermal velocity $u \approx (0.2\text{-}0.5)vT_e$ (see Section 4 in Chapter VI). Figure 49 shows the magnetic-field dependence of the total energy of the tail of accelerated hydrogen ions in the URAGAN-2 stellarator. These results show that the ion acceleration in the fields of microscopic fluctuations in the frequency range

---

*V. I. Sotnikov, V. D. Shapiro, and V. I. Shevchenko, Fiz. Plazmy 4, 450-459 (1978) [Sov. J. Plasma Phys. 4, 252 (1978)].

$\omega \sim \omega_{pi} \sim \omega_L$ in the plasma of a turbulent discharge is intensified as the magnetic field is increased. In a strong magnetic field the total energy of the high-energy ions approaches the energy density of the thermal electrons. It has been shown that this high-energy ion tail grows at a constant temperature of the bulk of the plasma ions and during a very strong perturbation of the plasma density by high-frequency noise, at $\omega \sim \omega_L$ and $2\omega_L$, which leads to the formation of high-frequency cavitons.[264,132] The modulational instability is thus regarded as one of the most probable causes of the observed intensification of the ion acceleration. This phenomenon probably results from an increase in the temporal and spatial correlation of the noise trapped in cavitons with increasing magnetic field.

CHAPTER V
ANOMALOUS CONDUCTIVITY AND TURBULENT HEATING IN TOROIDAL SYSTEMS

1 *REASONS FOR THE ANOMALOUS PLASMA CONDUCTIVITY IN TOROIDAL SYSTEMS*

Before we analyze the data from experiments on the anomalous resistance of plasmas and turbulent plasma heating we should examine the factors which can give rise to this anomalous behavior, working in the approximations of ordinary Coulomb interactions and of collective interactions. We will begin with Coulomb collisions. The most important effects here are (1) the collisions of electrons with impurity ions with $Z > 1$, (2) the electron runaway, and (3) the trapping of electrons at nonuniformities of the magnetic field of the confinement system.

Let us consider in more detail the changes which should be caused in the plasma resistivity η by each of these mechanisms.

When a hydrogen plasma is contaminated with impurities the effective charge

$$Z_{\text{eff}} = \frac{\sum_i n_i Z_i^2}{n_e}$$

may become considerably greater than one. Since $\eta \sim Z_{\text{eff}}/T^{3/2}$, the appearance of impurities in the plasma should raise its resistance in proportion to Z_{eff}. This effect can be important in both straight and toroidal discharges. It should be particularly strong in the very long-pulse discharges which occur in tokamaks, since the impurity pinch effect[41] should promote a rapid increase in Z_{eff}. The influence of this effect on the anomalous resistance has been studied in detail numerically[119] for the TM-3 tokamak. Those results yield $\eta/\eta_{coll} = 10$ as the greatest degree of anomalous behavior which can be ascribed to impurities without violating the quasineutrality condition.

Electron runaway should be more important in toroidal discharges with currents, since the high-energy part of the electron energy distribution can be caught up in acceleration considerably more easily in a torus. Furthermore, while the energy which the electrons in a straight discharge can acquire is set by the potential difference between the electrodes, in a toroidal confinement system the electrons can acquire very high energies in the loop electric field. The electron runaway should tend to reduce η. According to the theory of Ref. 208 this decrease in the resistance will become more pronounced with increasing electric field in the plasma.

The trapping of electrons at magnetic field nonuniformities should be most important for toroidal systems, which definitely have such nonuniformities. At low collision rates, for example, a derivation incorporating electron trapping yields[61]

$$\eta = \eta_{coll} \frac{1+0.35\varepsilon^{1/2}}{1-1.4\varepsilon^{1/2}}.$$

For intermediate collision rates the result is

$$\eta = \eta_{coll}[1+1.5\pi^{1/2}v_{Te}|\Theta|\varepsilon^2/\nu_{ei}].$$

An evaluation of this effect for the existing toroidal systems puts the degree of anomalous behavior at $\eta/\eta_{coll} \lesssim 5$.

It follows from the experimental data discussed above that the anomalous resistance in electric fields $E > E_{Dr}$ results from the driving of microscopic instabilities in a plasma with a current and reaches a value $\eta/\eta_{coll} \approx 10^3-10^4$.

The theory presently available suggests several instabilities which might be used as starting points in attempts to explain and classify the experimental data available. The most important of the instabilities which occur when the current flows in the direction along the magnetic field are, as mentioned earlier, (1) the Buneman instability,[42,201] (2) the ion acoustic instability,[66] and (3) the Drummond–Rosenbluth instability.[210]

Most of the measurements in the experiments discussed here corresponded to the conditions $v_{Te} > u > v_S$ and $T_e > T_i$, under which the ion acoustic and Drummond–Rosenbluth instabilities are the most likely. The growth rate of the latter instability, however, is small $\left(\sim \omega_{ci} \frac{u}{v_{Te}}\right)$, so that this instability should not cause any significant resistance anomaly.

The Buneman instability has occurred in very fast discharges, in which the condition $T_e \lesssim T_i$ was met in the preexisting plasma. This instability causes a rapid heating of the plasma electrons and leads to a situation with $T_e \gg T_i$. This instability should thus convert into the ion acoustic instability very quickly (in a time $\sim 1/(\omega_{pi})$.

In addition, there are current-dissipation mechanisms which involve the relatively weak streams of accelerated electrons in closed confinement systems with a strong magnetic field $H > (4\pi m_e c^2 n)^{\frac{1}{2}}$.

Kadomtsev and Pogutse[110] analyzed the transition to

an isotropic distribution for the accelerated electrons in a discharge with a current through the excitation of a cyclotron-harmonic resonance, taking the Doppler effect into account, $\omega - l^{*}\omega_{ce} - k_z v_z = 0$. They showed that under the condition $l^{*} < 0$ (corresponding to the anomalous Doppler effect) this mechanism can contribute significantly to the plasma pressure and can cause the measured conductivity σ to deviate markedly from that calculated from the "dimagnetic" temperature, σ_{coll}, under the assumption of Coulomb collisions alone. The anomalous conductivity calculated under this assumption is

$$\frac{\sigma_{coll}}{\sigma} = \frac{L^{-3/2}}{1+\Gamma}\left(\frac{E}{E_{Dr}}\right)^{3/2}.$$

where $L = E/E'_{Dr}$, $\Gamma = \tau_{ei}E/(L^3\tau^{*}_{E}E_{Dr})$, τ^{*}_{E} is the lifetime of fast particles, and E'_{dr} is the critical electric field calculated from the temperature determined from the conductivity. For $\Gamma < 1$ there should be a dependence $\sigma_{coll}/\sigma \sim (E/E_{Dr})^{3/2}$, while for $\Gamma > 1$ we should have $\sigma_{coll}/\sigma \sim (E/E_{Dr})^{\frac{1}{2}}$. When the particular features of the fast-particle motion in the nonuniform toroidal magnetic field are incorporated in this model, the dependence $\sigma_{Coll}/\sigma = c^{*}(\omega_{ce}/\omega_{pe})^{3}$ is found, where c^{*} is a coefficient of order unity.

Rudakov[148] has studied the appearance of an anomalous electrical conductivity and of turbulent plasma heating in a strong magnetic field as a consequence of the joint effects of ion acoustic and cyclotron instabilities.

Vekshteǐn and Sagdeev[58] examined the weak-turbulence equation for the current; this case has an exact solution, in the one-dimensional case. This case can be quite pertinent to a highly magnetized plasma, in which the energy lifetime of the plasma particles is very long, since it is essentially impossible for electrons to move in the direction across the magnetic field. In this case the electron distribution function changes in a self-similar way a short time after the imposition of the electric field. In this case we would have $W/nT_e \sim t^{-1}$ and a current velocity $u = \alpha^* \frac{eE}{m_e} t$, where $\alpha^* = 1$ (Ref. 58).

The self-similar equation for the quasilinear case in the one-dimensional formulation thus yields a solution describing a free acceleration of plasma electrons.

In several cases, the theory which has been worked out regarding possible mechanisms for the anomalous electrical conductivity has led to a broader view of the experimental results which have been accumulated over nearly two decades. The theory also makes it possible to classify the results in a definite way.

2 EXPERIMENTS ON THE ANOMALOUS RESISTANCE AND THE TURBULENT CURRENT HEATING OF PLASMAS IN TOROIDAL SYSTEMS

The chronological order in which the most important papers on the anomalous resistance and turbulent plasma heating in tori appeared was outlined in Chapter I. We can thus limit the present section to an analysis of the most general behavior which has emerged and a discussion of the results of this analysis.

All the toroidal devices which have been used to study this problem may be classified as either (1) devices constructed especially for a study of turbulent plasma heating by currents, most of which have relatively weak magnetic fields and large-amplitude electric fields ($E \lesssim 100$ V/cm), induced in the plasma column by rf pulse circuits with a frequency ~1 MHz, or (2) toroidal devices intended primarily for a study of the thermal insulation of the plasma, so that the magnetic fields are quite high and the electric fields low ($E \lesssim 1$ V/cm). The electric fields in this second case are induced by pulses ranging in length from a few hundred microseconds to a few milliseconds.

For all the devices of the first group, the experiments were planned under the assumption that it would be necessary to use electric fields much stronger than the Dreicer critical field. To a large extent this approach was a reaction to unsuccessful attempts to accelerate electrons in so-called plasma betatrons. The failure was ascribed to beam instabilities. In the very first experiments in the devices of this first group it was found that the current discharge behaves in a very peculiar way at high electric fields in the plasma. This peculiarity is easily seen on a series of oscilloscope traces (Fig. 50) obtained in a toroidal experiment in the VIKhR'-1 device.[167] It follows from these traces that the current in the plasma stops flowing long before the longitudinal electric field vanishes. Measurements with magnetic probes and image-converter streak cameras did not reveal a destruction of the plasma column or a loss of plasma to the wall of the vacuum chamber. The resistance of the plasma turned out to be anomalously high, reaching 30 Ω during the first current half-period for the particular conditions of this experiment.

The time evolution of the resistance can be illustrated by the results in Fig. 51, from Ref. 23. Figure 52, from Ref. 72, shows the dependence of the discharge current on the electric field. Elementary calculations for toroidal discharges of this type put the plasma resistance at a level roughly two orders of magnitude greater than the

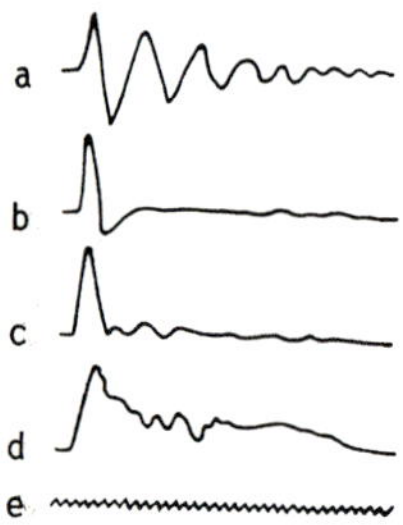

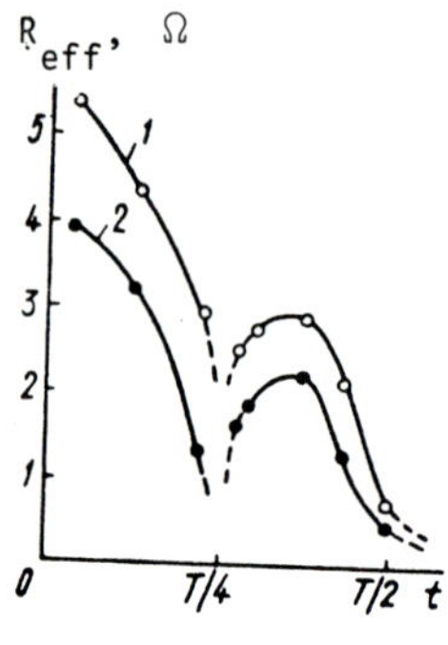

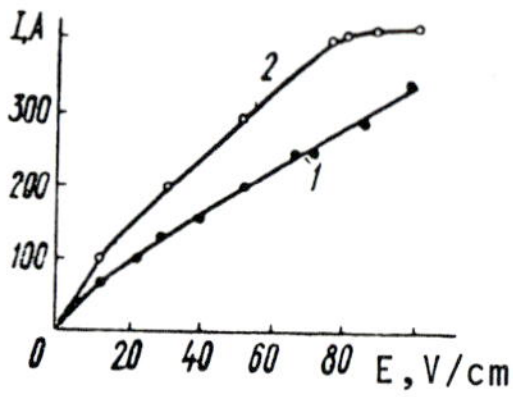

FIGURE 50 *a*) Electric field; *b*) current in the plasma; *c*) hard component of the x-ray emission; *d*) emission in the Hα line; *e*) calibration signal, $\omega/2\pi$ = 12 MHz. $n = 1 \cdot 10^{12}$ cm^{-3}; E = 150 V/cm; H = 1 kOe (Ref. 167).

FIGURE 51 Time evolution of the resistance of the plasma secondary for various loop voltages. *1*) E_{max} = 250 V/cm; *2*) E_{max} = 200 V/cm. $n = 3 \cdot 10^{13}$ cm^{-3}; H_0 = 5 kOe.

FIGURE 52 Dependence of the discharge current on the electric field. *1*) First current maximum; *2*) second maximum. $n = 1 \cdot 10^{12}$ cm^{-3}; H_0 kOe.

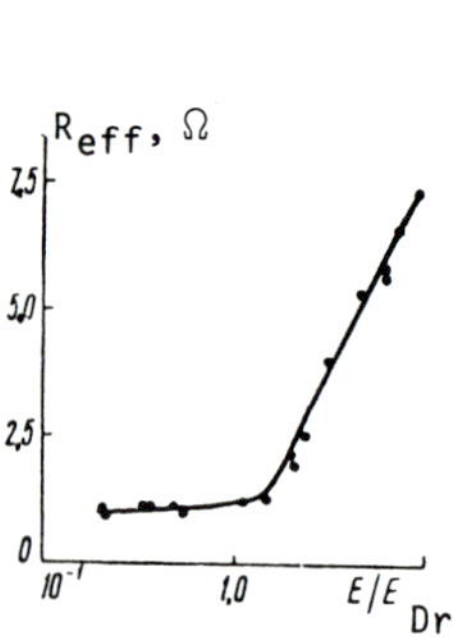

FIGURE 53 Dependence of the effective plasma resistance on the ratio of the applied electric field to the Dreicer field. $n = 2 \cdot 10^{13}$ cm^{-3}; $E \lesssim 20$ V/cm; H_0 = 8 kOe.

ohmic resistance of the cool plasma produced by the preionization.

The resistance of the plasma column was measured at various electric fields in the SIRIUS stellarator.[43] The approach used to produce rather strong electric fields, $E \approx 10\text{-}20$ V/cm, was similar to that used in small toroidal devices. Figure 53 shows the dependence of the effective plasma resistance on the ratio of the applied electric field to the Dreicer field. This dependence was subsequently confirmed in experiments in the VIKhR'[76] and TWIST devices.[221,222] The measurements in the SIRIUS stellarator were supplemented with measurements at electric fields in the range $E \approx (10^{-2}\text{-}1)E_{Dr}$ (Refs. 47 and 48). When all these results on the E dependence of the plasma resistance are combined, one obtains a broad view of the behavior of the plasma conductivity as a function of the electric field. The dependence also has been studied over the entire field range in a single device[45] (Fig. 54). At $E/E_{Dr} < 2\cdot 10^{-2}$ the plasma conductivity increases with increasing electric field. This part of the curve corresponds to classical ohmic heating. At $E/E_{Dr} > 2\cdot 10^{-2}$ the conductivity drops rapidly, and it reaches a plateau at $E \approx 0.2\,E_{Dr}$ changing only slightly up to $E \approx 1.5\,E_{Dr}$. With a further increase in the field there is yet another drop in the conductivity. The limitation on σ observed in this experiment signified the onset of an effective frictional force, which imposed a limit on the electron drift velocity $u = \sigma E/en$. Estimates of this limit based on the results of many experiments yielded $u \sim v_S$.

All these measurements, we should note, were carried out in relatively weak magnetic fields, under the condition

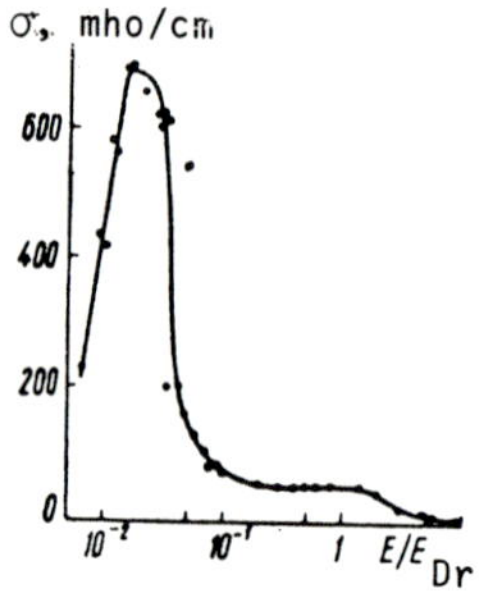

FIGURE 54 Dependence of the plasma conductivity on the ratio of the applied electric field to the Drecier field. $n = 2 \cdot 10^{13}$ cm^{-3}; $E \lesssim 20$ V/cm; $H_0 = 8$ kOe.

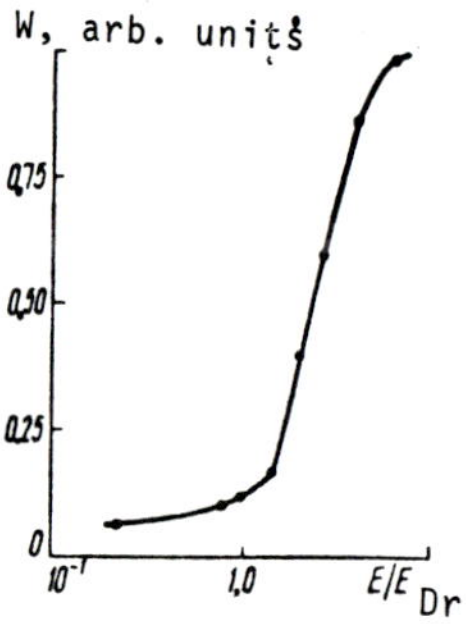

FIGURE 55 Dependence of the emission intensity at the wavelength $\lambda = 3$ cm on the ratio E/E_{Dr}. $n = 2 \cdot 10^{13}$ cm^{-3}; $E \lesssim 20$ V/cm; $H_0 = 8$ kOe;

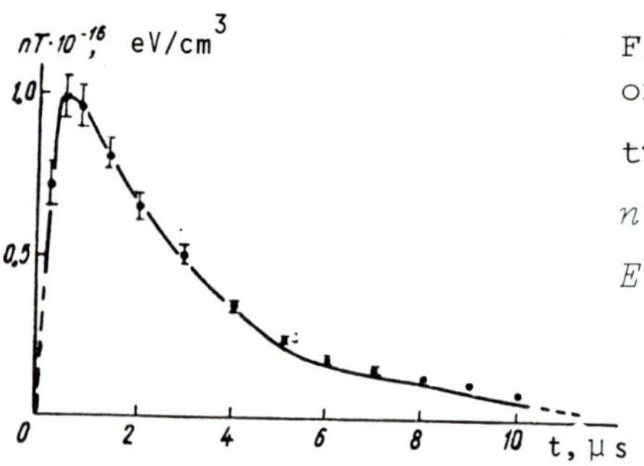

FIGURE 56 Time evolution of the plasma energy during turbulent heating.[23] $n = 3 \cdot 10^{13}$ cm^{-3}; $H_0 = 5$ kOe; $E = 250$ V/cm.

$$\frac{\omega_{ce}}{\omega_{pe}} < 1.$$

Estimates of the effects of the binary-collision mechanisms for a limitation on σ discussed in Section 1 showed that these mechanisms could not explain the observed effects. Subsequent experiments were thus aimed primarily at identifying the reasons for the anomalous conductivity. The effort naturally concentrated on a study of plasma waves, particularly measurements of the microwave emission. It was found that at electric fields above the threshold $E_{thr} \approx 10^{-2} E_{Dr}$ the plasma of a toroidal discharge with a current becomes a source of nonthermal microwave emission at an intensity four to six orders of magnitude above the thermal level.[72,74] In the case $E < E_{Dr}$ the emission is most intense near the plasma frequency ω_{pe}, while at $E \gtrsim E_{Dr}$ the emission occurs over a broad spectral range from ω_{pi} to ω_{pe}. Measurements[45] of the E/E_{Dr} dependence of the microwave emission power revealed results (in a certain wavelength interval) in qualitative accordance with the change in $R_{eff}(E/E_{Dr})$ (Fig. 55). This experimental fact directly confirmed the suggestion that the superthermal microwave emission and the anomalous resistance were interrelated. A study of the waves in the plasma revealed that the energy density of the waves near the electron plasma frequency was of the same order of magnitude as that near the ion plasma frequency.[75] The discovery and study of Raman (or combinational) scattering of electromagnetic waves by plasma waves (or Langmuir waves) in plasmas made it possible to estimate the turbulent noise level W/nT and to show that this noise level was several orders of magnitude above the thermodynamic-equilibrium value.[74,141]

The microwave activity of plasmas will be discussed in detail below.

A second effect confirming that plasma turbulence affects the electric-field dependence of the plasma resistance was the emission of x radiation over a broad spectrum by the plasma of an electric discharge.

Spectral study of the soft x-ray emission in the vacuum chamber led to the conclusion that a plasma in a strong electric field contains a group of electrons with a nearly Maxwellian velocity distribution, with a "temperature" ranging from 1 to several keV (Refs. 44 and 169). These experiments showed that when collective interactions are driven by a current in a plasma a significant fraction of the plasma electrons are heated. Diamagnetic measurements made it possible to follow the dynamics of the heating process itself and to study this process quantitatively.[23,214] Figure 56 shows the change in the energy of the plasma during the turbulent heating in a small tokamak in which a strong electric field was imposed on a quasisteady current by an rf circuit. It can be seen that the plasma was heated very rapidly (in a time on the order of 0.3 μs).

Estimates of the efficiency of the plasma heating in the various devices of this first group yielded values ranging from 4% to 90%. Measurements of the plasma energy nT at various electric field amplitudes in the plasma revealed $nT \sim E^2$ (Fig. 57). In a recent study in the T-5M device the electron temperature was measured by laser-scattering.[69] These measurements, and also the results of Ref. 242, completely confirm the earlier data from dimagnetic measurements on the effective heating of the plasma electrons.

A rapid non-Coulomb heating of the ions was observed in all the experiments on turbulent current heating of plasma in the devices of the first group. It was found that the ion distribution function has a tail which corresponds to a definitely collective acceleration of ions in the turbulent plasma. It should be pointed out that the non-Coulomb ion heating was observed in devices with approximately steady discharges (the ZETA[32] and the AL'FA[11,98]), at a time before the method of turbulent plasma heating had been suggested. In those experiments it was simply noted that the ion temperature in the current-carrying plasma is directly proportional to the ion charge Z (Fig. 58). Since the mechanism for ion heating in a turbulent plasma with a current has been discussed in more detail previously, we will not pursue this question here.

Let us examine some of the results of the research in toroidal devices with current heating with strong electric fields, $E > E_{Dr}$, and weak magnetic fields, $\omega_{ce}/\omega_{pe} < 1$.

1. During the current-heating stage, with strong electric fields, the impedance of the plasma column is a pure resistance and is far higher than that calculated from the theory of binary Coulomb collisions.

2. The E dependence of the plasma resistance correlates with the E dependence of the microwave emission intensity.

3. The plasma conductivity becomes anomalous at a threshold electric field in the plasma, $E_{thr} \approx 2 \times 10^{-2} E_{Dr}$.

4. The dependence $\sigma(E/E_{Dr})$ has a clearly defined plateau.

5. The electron drift velocity stabilizes at a level

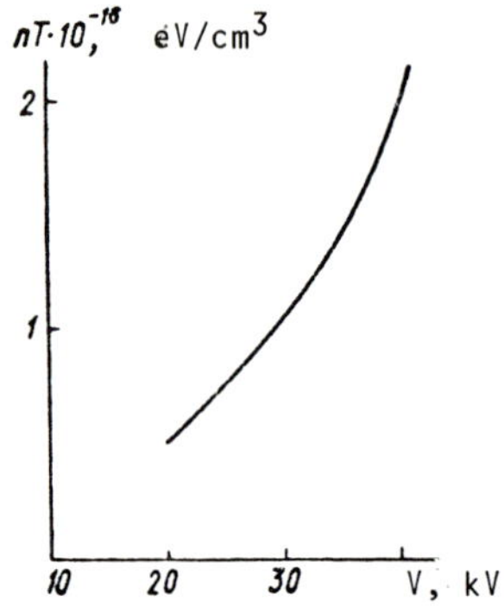

FIGURE 57 The plasma energy nT versus the voltage to which the capacitor bank of the pulse circuit is charged. $n = 3 \cdot 10^{13}$ cm^{-3}; $H_0 = 5$ kOe (Ref. 23).

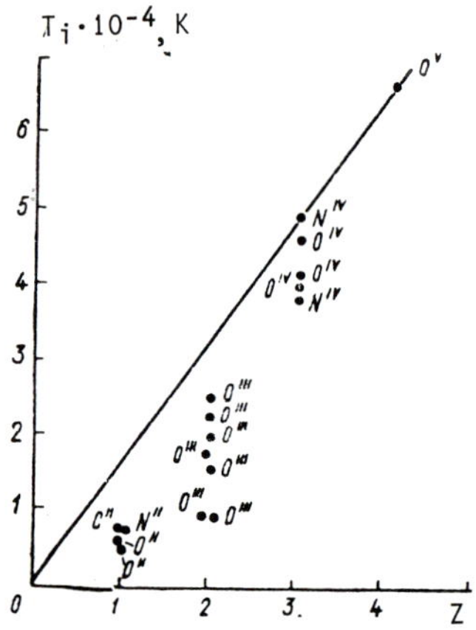

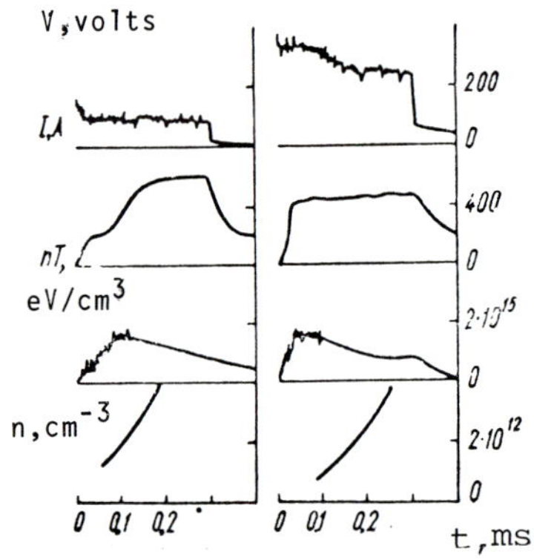

FIGURE 58 Z dependence of the impurity ion temperature in the AL'FA device.[97]

FIGURE 59 Oscilloscope traces of the voltage V, the discharge current I, the signal from the diamagnetic probe, nT, and the plasma density n (H_0 = 16 kOe).

$u \sim v_S$ during turbulent heating.

6. Intense waves are observed near the frequencies ω_{pi} and ω_{pe} during turbulent current heating.

7. While the resistance is anomalous, the efficiency of the heating, of both electrons and ions, increases sharply.

We turn now to an analysis of the experimental data obtained in closed magnetic confinement systems with approximately steady discharges in strong magnetic fields, with $\omega_{ce}/\omega_{pe}>1$. Figure 59 shows two series of oscilloscope traces, obtained under identical amplification conditions, which demonstrate the time evolution of the torus loop voltage, the plasma current, and the longitudinal magnetic flux due to the plasma diamagnetism. It is easy to see that as the electric field in the plasma is increased the plasma heating rate (the rate of increase of nT) increases, while the current remains essentially "frozen." The plasma temperature inferred from the dimagnetic effect is much higher than the temperature found from the conductivity measurements. Figure 60 shows the E/E_{Dr} dependence of the ratio of the measured plasma temperature to that calculated from the conductivity, for small values of E/E_{Dr}. This dependence was obtained by two different methods (dimagnetic measurements and laser scattering) in the Model C* and SIRIUS stellarators.[46] At $E < 2\cdot10^{-2}E_{\mathrm{Dr}}$ the plasma temperature agrees with that calculated from the conductivity, while at $E > 2\cdot10^{-2}E_{\mathrm{Dr}}$ there is a discrepancy; the discrepancy increases with increasing E/E_{Dr}. As T_D/T_σ increases there is an increase in the resistance of the plasma column. Measurements of the current at fixed values of the density for various torus loop voltages

*Dimock D. and E. Mazzucato, *Phys. Rev. Lett.* 20, 713-714 (1968).

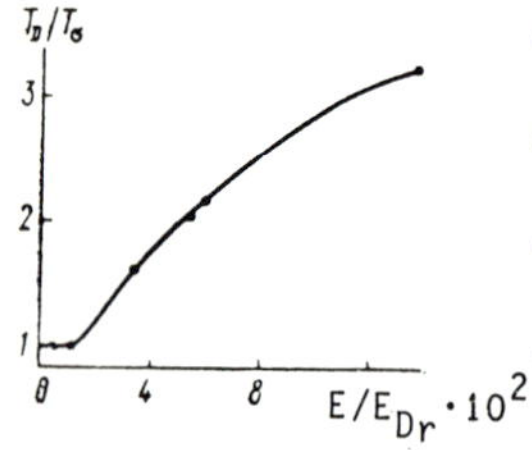

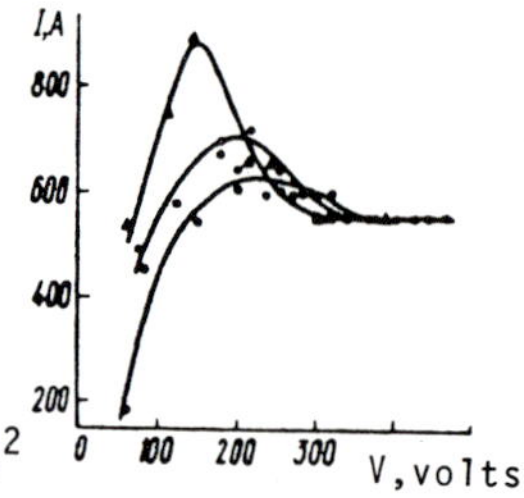

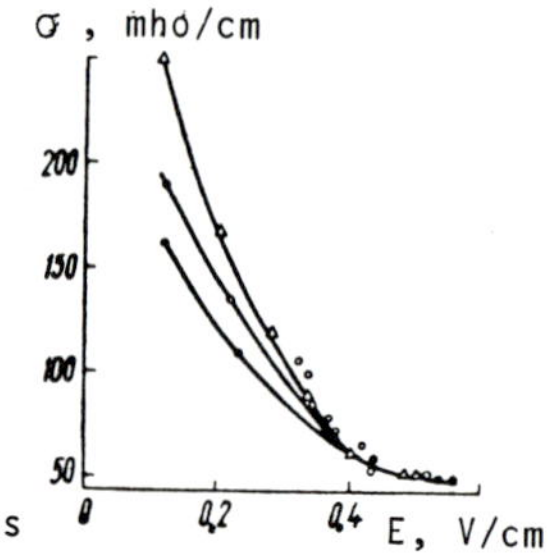

FIGURE 60 The ratio of the "diamagnetic temperature" to the temperature calculated from the conductivity versus E/E_{Dr} (Refs. 47 and 48).

FIGURE 61 Discharge current versus the loop voltage in the SIRIUS stellarator (H_0 = 16 kOe). Δ) $n = 4\cdot 10^{12}$; o) $n = 3\cdot 10^{12}$ cm^{-3}; ●) $n = 2\cdot 10^{12}$ cm^{-3}.

FIGURE 62 Plasma conductivity versus the electric field (H_0 = 16 kOe). Δ) $n = 4\cdot 10^{12}$ cm^{-3}; o) $n = 3\cdot 10^{12}$ cm^{-3}; ●) $n = 2\cdot 10^{12}$ cm^{-3}.

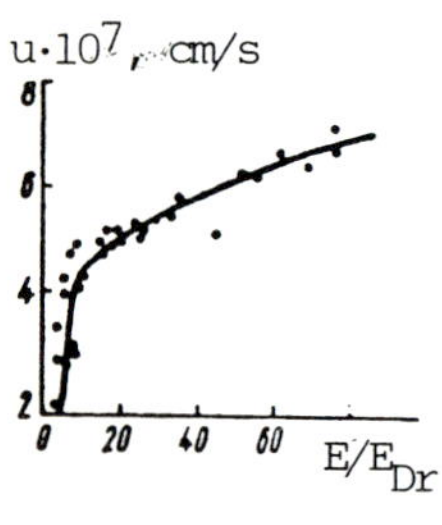

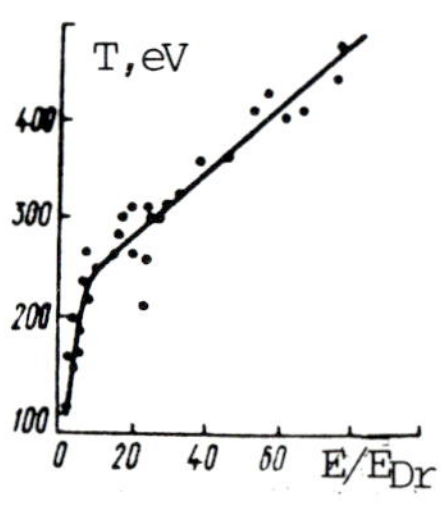

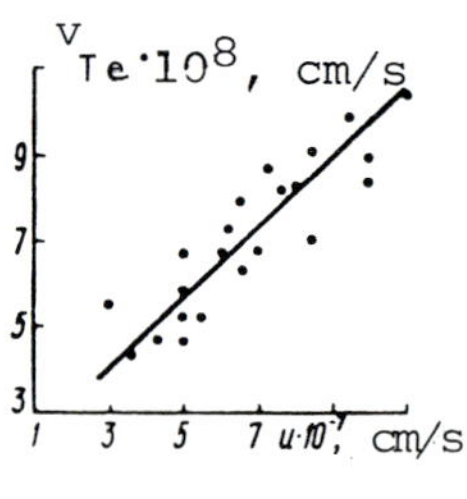

FIGURE 63 The electron drift velocity versus E/E_{Dr}. H_0 = 16 kOe; n = (2-4) x 10^{12} cm^{-3}.

FIGURE 64 Plasma temperature versus E/E_{Dr}. H = 16 kOe; n = (2-4) x 10^{12} cm^{-3}.

FIGURE 65 Relationship between the electron thermal velocity and the electron drift velocity in a magnetized plasma.

revealed that the current initially increases and then begins to decrease; as the electric field is raised further, the current tends toward saturation (Fig. 61). Figure 62 shows the dependence of the plasma conductivity on the electric field in the plasma. We see that σ falls off with increasing E and that at $E > 0.2$ V/cm the behavior is $\sigma \sim 1/E$.

In weak electric fields the electron drift velocity increases rapidly with increasing E to a certain point, beyond which the increase in u slows down (Fig. 63). The slope change on the $T_e(E/E_{Dr})$ curve is typically observed at the same fields as that on the $u(E/E_{Dr})$ curve (Fig. 64). From these figures we immediately find the dependence $u \sim v_{Te}$ shown in Fig. 65 (Ref. 49). It turns out that in electric fields $E > E_{Dr}$ the current velocity in a plasma with $\omega_{ce}/\omega_{pe} > 1$ stabilizes at $u \approx (10-15)v_S$, i.e., at a level higher than in weak magnetic fields. In weaker electric fields, $\alpha = u/v_s$ varies with E.

Measurements taken in quasisteady discharges revealed the electric-field dependence of the anomalous plasma conductivity. Figure 66 illustrates the results with the ratio of the conductivity calculated under the assumption of Coulomb collisions to the measured conductivity, plotted as a function of E/E_{Dr}, according to experiments carried out in the SIRIUS stellarator. Clearly, the dependence is linear. At fields $E/E_{Dr} < 1$, however, it is described approximately by $\dfrac{\sigma_{coll}}{\sigma} \sim \left(\dfrac{E}{E_{Dr}}\right)^{1/2}$ (see Fig. 67, which shows results obtained from the Model C, SIRIUS, and TM-3 devices[50]).

It should be noted that the cases $E > E_{Dr}$ and $E < E_{Dr}$ in plasmas in a strong magnetic field differ in yet another regard. In the case $E < E_{Dr}$ in a plasma with a current

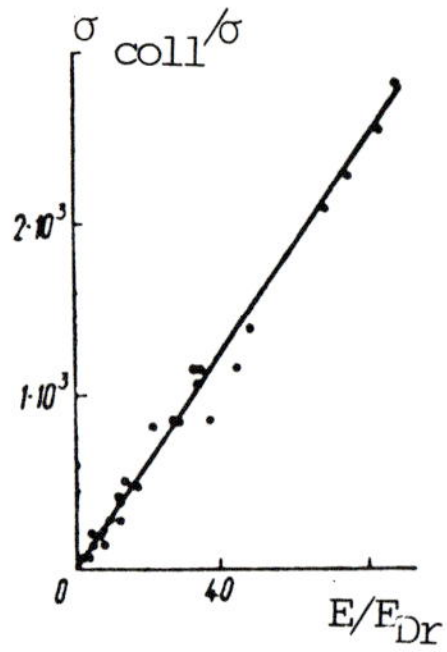

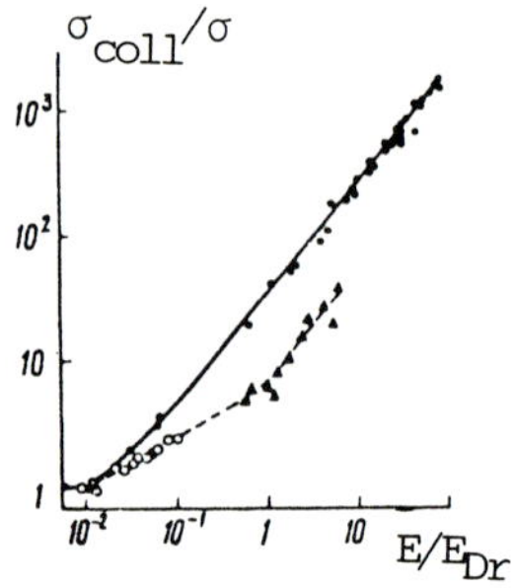

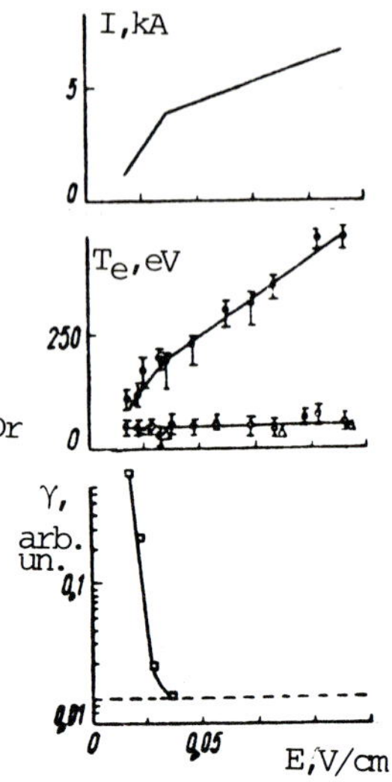

FIGURE 66 The conductivity anomaly versus E/E_{Dr}. H = 16 kOe; n = (2-4) x 10^{12} cm^{-3} (Ref. 49).

FIGURE 67 The conductivity anomaly versus E/E_{Dr}. •) Data from the SIRIUS stellarator; o) data from the Model C stellarator; Δ) data from the TM-3 tokamak.

FIGURE 68 Dependence of various properties on the electric field. Solid lines) Total current; Δ) electron temperature according to laser scattering and the microwave emission (o: "red" satellite; •: "violet" satellite); □) hard x-ray emission from the chamber wall in a hydrogen discharge in the URAGAN-2 stellarator ($n = 6 \cdot 10^{12}$ cm^{-3}; H = 7.8 kOe; $\omega_{ce}/\omega_{pe} \cong 1$); Dashed line) energy level of 10-20 keV.

the electrons are accelerated to high energies, as demonstrated by x-ray emission from the chamber wall. In the case $E/E_{Dr} > 1$ this electron acceleration is not observed.

Figure 68 shows the field dependence of the total current and of the electron temperature according to laser scattering and the microwave emission near the frequencies $\omega_{pe} \approx \omega_{ce}$. Also shown here is the intensity of the hard x-ray emission from the chamber wall for a purely hydrogen discharge in the URAGAN-2 stellarator. These results were obtained at a constant plasma density, $n \simeq 6\cdot 10\ \mathrm{cm}^{-3}$, in a field H = 7.8 kOe, with a corresponding ratio $\omega_{ce}/\omega_{pe} \simeq 1$. Near $E_{\mathrm{thr}} \simeq 0.03$ V/cm there is a change in the slope on the voltage—current characteristic, and there is an effective deceleration of the fast electrons. The spectrum of the thermal microwave emission near $\omega_{pe} \approx \omega_{ce}$ reveals a splitting into two satellites: a "red" satellite $\omega_{\mathrm{red}} \approx \omega_{pe} - 2\omega_{pi}$, and a "violet" one $\omega_{\mathrm{viol}} \simeq \omega_{pe} + 4\omega_{pi}$. The intensities of these satellites vary in different ways with the electric field. Measurements of the intensity of the red satellite reveal a temperature dependence which is the same as that of the transverse electron temperature, determined from laser scattering. Again in the case $\omega_{ce} > \omega_{pe}$, as in the devices of the first group, it was found that when the resistance is anomalous the plasma is a source of waves near ω_{pe} and ω_{pi}.

The experimental work on turbulent discharges with currents in toroidal devices with strong magnetic fields, $\omega_{ce} > \omega_{pe}$, thus revealed the following.

1. At electric fields $E > E_{\mathrm{Dr}}$ there is a linear relationship between the current velocity and the thermal velocity of the plasma electrons.

2. At $E > E_{\mathrm{Dr}}$ the electron drift velocity is "frozen" at a level of $(10\text{-}15)v_S$.

3. The ratio of the conductivity calculated from the measured plasma temperature under the assumption of purely Coulomb collisions to the experimental value, $\sigma_{\mathrm{coll}}/\sigma$, is directly proportional to E/E_{Dr} at $E > E_{\mathrm{Dr}}$; at lower values of E, the dependence $\sigma_{\mathrm{coll}}/\sigma \sim (E/E_{\mathrm{Dr}})^{1/2}$ is observed.

4. At the field $E = E_{\mathrm{Dr}}$ in a longitudinal current discharge in a strong magnetic field, the electron acceler-

ation observed at $E < E_{Dr}$ stops.

Comparison of all these experimental results with the theory discussed in Chapter III leads to the conclusion that in both weak and strong magnetic fields the anomalous plasma resistance in a toroidal current discharge with $E > E_{Dr}$ results from an ion acoustic instability. It should be noted, however, that the level at which the current velocity stabilizes depends on the magnetic field. This dependence is not embodied in the theory, which deals with only the limiting cases $\omega_{ce}/\omega_{pe} \ll 1$ and $\omega_{ce}/\omega_{pe} \gg 1$. Figure 69 shows the experimental results on the dependence $u/v_{Te}(\omega_{ce}/\omega_{pe})$. While we find $u \simeq v_S$ in weak magnetic fields, in strong fields we find $u \simeq 10v_S$ and $u/v_{Te} \sim (\omega_{ce}/\omega_{pe})^{1/3}$. The change in slope on this curve was determined from experiments in the GROM-3. Recent experiments in the VINT-20, SIRIUS, and URAGAN-2 devices have shown that the slope change in toroidal systems should correspond to a value $\omega_{ce}/\omega_{pe} \simeq 1$.

It is natural to suggest that the macroscopic relationships found in each particular case should depend on the particular type of energy loss from the plasma, especially for quasisteady toroidal discharges. Let us examine this question.

We consider yet another experiment, helpful for generalizing the data obtained in a strong magnetic field. Measurements of the turbulent noise level in a current-carrying plasma from the emission of forbidden lines of helium have yielded the dependence $(W/nT)(1/\omega_{ce}) \sim 1/\sigma$, shown in Fig. 70 (Ref. 28). From this dependence it follows that

$$\nu_{eff} \sim \omega_{pe} \cdot \frac{W}{nT} \cdot \left(\frac{\omega_{pe}}{\omega_{ce}}\right),$$

where ν_{eff} is the effective collision rate. The quasilinear theory for the ion acoustic instability predicts

$$\nu_{eff} \sim \omega_{pe} \, (W/nT)(v_{Te}/v)^3.$$

If we assume that the velocity of the resonant particles is $v \sim u$ and use the experimental result $u/v_{Te} \sim (\omega_{ce}/\omega_{pe})^{1/3}$, we immediately find $\nu_{eff} \sim 1/H$, which is observed experimentally. In turn, this result means that as the magnetic field is increased the effective collision rate should fall off, and as $H \to \infty$ we should find $v_{eff} \to 0$. The phenomenological picture found experimentally is thus consistent with the self-similar solution discussed in Ref. 58.

Pertinent in this connnection are some studies by Al'bikov et al.[2] and Ivanov et al.,[104] who showed that under certain conditions [with a substantial lowering of the density ($n \lesssim 10^{11}$ cm^{-3}) and a rather strong magnetic field (H = 7 kOe)] all the plasma electrons are accelerated to energies of about 0.5 MeV in a system in which strong electric fields are used. We can begin to understand this result, which appeared to completely refute all the arguments regarding the onset of a beam-plasma instability, only by working from the phenomenology discussed above, since in this experiment the condition $\omega_{ce}/\omega_{pe} \gg 1$ held.

Before we analyze the heat loss in toroidal turbulent current discharges, we will summarize the results observed experimentally in the various heating regimes.

Regime 1: $E \gtrsim E_{Dr}$, $\alpha = u/v_s \sim 1$, $\omega_{ce}/\omega_{pe} < 1$, σ = const

Regime 2: $E \lesssim E_{Dr}$, $\alpha = \alpha(E/E_{Dr}, \omega_{ce}/\omega_{pe})$, $\omega_{ce}/\omega_{pe} \simeq 1$, $\sigma \sim 1/E$.

Regime 3: $E > E_{Dr}$, $\alpha \sim (\omega_{ce}/\omega_{pe})^{1/3}$, $\omega_{ce}/\omega_{pe} > 1$, $\sigma \sim 1/E$.

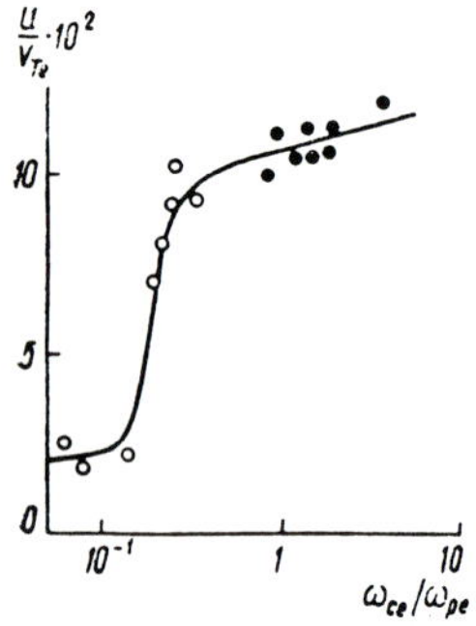

FIGURE 69 Dependence of the level at which the drift velocity stabilizes on the ratio ω_{ce}/ω_{pe}. o) Data from the GROM-3 device; ●) the SIRIUS stellarator.

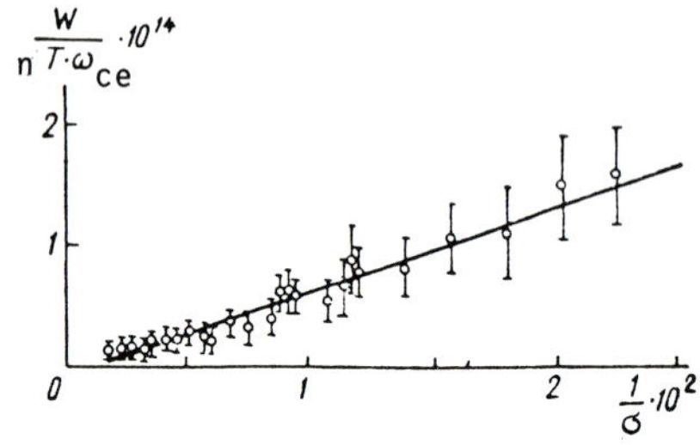

FIGURE 70 The dependence

$$\frac{W}{nT}\frac{1}{\omega_{ce}} \sim \frac{1}{\sigma}.$$

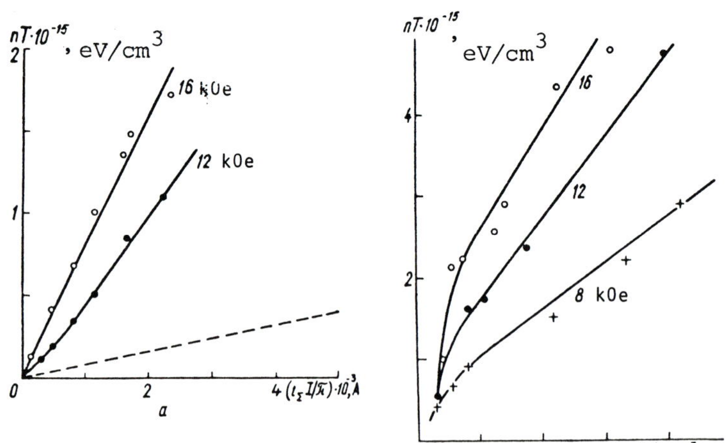

FIGURE 71 The functional dependence $nT(\ell_{\Sigma}I/\pi)$: *a)* For $E < E_{Dr}$; *b)* For $E > E_{Dr}$ in a plasma with $\omega_{ce}/\omega_{pe} \gg 1$.

3 PLASMA TRANSPORT DURING TURBULENT HEATING IN TOROIDAL SYSTEMS

It was shown in Sections 1 and 2 that the dependence $\sigma(E/E_{\mathrm{Dr}})$ which emerges from experiments in devices with weak magnetic fields and strong electric fields is similar to that observed in high-current straight discharges. Furthermore, the condition $\beta_{\varphi} \simeq 1$ is violated even during the heating in most experiments, so that the magnetic configuration should be distorted, and magnetic lines of force should go outside the confinement volume to the chamber wall. In this case a toroidal system becomes completely analogous to a straight system, since the heat is transferred by the hot group of electrons directly along the magnetic lines of force.

Research on the thermal insulation of plasmas in toroidal confinement devices with strong magnetic fields and with approximately steady current heating has shown that the energy loss across the magnetic field is anomalously high in comparison with that calculated from the neoclassical transport theory and is determined by the electron thermal conductivity. In particular, in tokamaks and stellarators at moderate values of the ohmic-heating current, the electron thermal conductivity is determined by the pseudoclassical expression $\chi_e = c^2\rho^2_{\varphi}\nu_{\mathrm{eff}}$, where the effective collision rate is calculated from the measured electrical conductivity of the plasma. The empirical coefficient c in this expression has been found to lie between 1 and 3 for all devices.

In order to distinguish between regimes 2 and 3 as defined at the end of the preceding section, we consider some experimental results from Refs. 297 and 197, obtained in a stellarator with $E \gtrless E_{Dr}$, $\omega_{ce}/\omega_{pe} \gtrsim 1$ and $E > E_{Dr}$, $\omega_{ce}/\omega_{pe} > 1$. Another distinction between these two regimes is that at $E \gtrless E_{Dr}$ an electron acceleration is observed in the plasma. At $E > E_{Dr}$, as we mentioned earlier, the acceleration is halted.

If we assume that the pseudoclassical expression is valid, then the relation $nT \sim I\iota_{\Sigma}H$ should hold. The total rotational transform ι_{Σ} for the stellarator must include the rotational transform caused by the current in the plasma and that caused by the helical windings. Figure 71 shows curves of $nT(I\iota_{\Sigma})$ found for the two cases listed above. The pseudoclassical law is satisfied in case a, i.e., in the case of relatively weak electric fields and an only slightly anomalous electrical conductivity ($\sigma_{coll}/\sigma \sim 10$) for this particular experiment. In the case of a very anomalous conductivity and at $E > E_{Dr}$ the pseudoclassical law does not describe the loss. Interestingly, when the pseudoclassical law $nT \sim I\iota_{\Sigma}H$ holds, the electron drift velocity exceeds the phase velocity of the drift waves, $u > \omega^*/k_{\parallel}$, where $\omega^* = ck_{\perp}T_e/eaH$ and $a = (d\ \ln n/dr)^{-1}$ (Fig. 72), and is proportional to it: $u \sim \omega^*/k$. The quantity $\alpha = u/v_S$, on the other hand, varies in a different way.

It has been suggested that the physical nature of the pseudoclassical behavior of the electron thermal conductivity is determined by a distortion of the electron velocity distribution caused by a drift instability in a discharge with a current. This effect can be taken into account qualitatively by making the assumption that the drift velocity u is stabilized at the level of the phase velocity of a drift wave. It is then a simple matter to work from the heat balance equation $\frac{3}{2} \cdot \frac{nT}{\tau_E} = \frac{j^2}{\sigma}$ and the assumption $k_{\parallel} = \frac{\iota}{L} = \frac{H_{\varphi}}{rH_0}$ to derive an expression

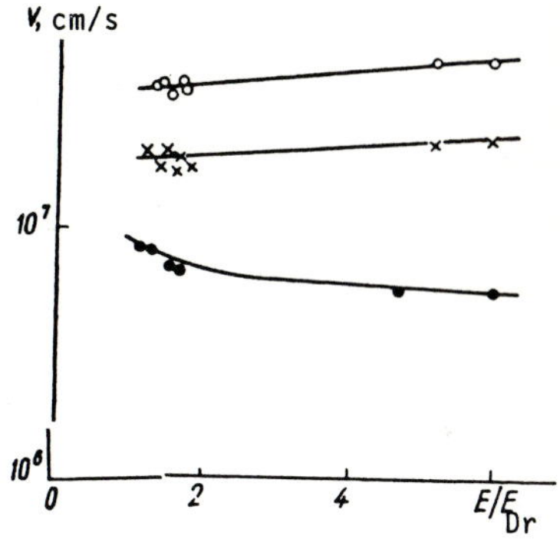

FIGURE 72 Behavior of u, v_S, and ω^*/k in a discharge with $E/E_{Dr} < 1$ and $\omega_{ce}/\omega_{pe} \lesssim 1$. x) $\omega/k_{\parallel}$; o) u; ●) v_S.

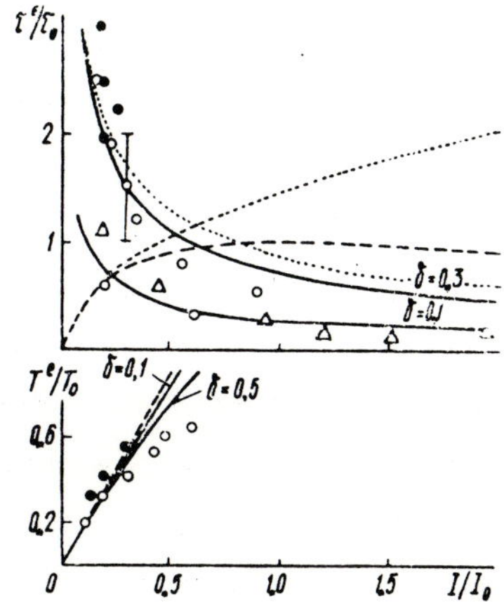

FIGURE 73 Normalized electron energy lifetime and normalized electron temperature vs. the ohmic-heating current for ion acoustic turbulence (solid curves) and for the classical Coulomb conductivity (dotted curves). ●) Electron energy lifetime without allowance for convective loss according to the URAGAN-1 data; o) URAGAN-2; △) CLEO.

for χ_e which corresponds to that proposed by Artsimovich. So far, this phenomenological picture of the pseudoclassical heat transfer, which is based on the experimentally observed proportionality $u \sim \frac{\omega^*}{k_\parallel}$ (which holds in a rather narrow interval of plasma parameters), has found nothing approaching a solid theoretical basis.

The question which is the most difficult to understand and which has received the least experimental study is that of the thermal conductivity of a plasma in the presence of a well-developed ion acoustic instability, under the condition $u/v_S \approx const.$ (i.e., regime 3).

Recent tokamak and stellarator experiments have shown that the pseudoclassical expression holds only at relatively low currents. As the ohmic-heating current is raised, there is a deviation from the pseudoclassical law in the direction of an increased heat loss. In these experiments the energy lifetimes of the plasma and the particle lifetimes typically become more nearly equal. In this case the heat loss due to diffusion of particles must be taken into account in an examination of the energy balance of the discharge. Figure 73 shows the normalized electron energy lifetime and the normalized electron temperature calculated for a Coulomb conductivity and for a turbulent ion acoustic conductivity, with and without heat loss due to diffusion. Shown for comparison here are experimental data obtained in the CLEO and URAGAN stellarators. Taken together, these results show that in discharges with currents in these devices the loss can be described under the assumption of a turbulent electrical conductivity caused by an ion acoustic instability, with a diffusive heat loss.

The mechanism for the heat loss in tokamaks with strong

magnetic fields during turbulent plasma heating was examined in detail by Rudakov.[148] He showed that under the conditions corresponding to an anomalous resistance the ion acoustic instability can drive an overheating instability, which in turn redistributes the current over the cross section of the plasma column. The calculations showed that if the electron energy loss is assumed to result from the diffusion of trapped particles, as a result of scattering by the electric fields of ion acoustic waves, then it follows from the energy balance equation that the heat is dissipated primarily in that tube having the higher temperature. This effect should cause a redistribution of the current, whose density will increase in the region with the elevated temperature. The minimum size of this tube is determined by equating the heating rate to the rate of heat loss:

$$\frac{r_0}{r} = \beta\eta^{1/2}, \quad \beta = \frac{8\pi n T_e}{H_\varphi^2}, \quad \eta = \left(\frac{r}{R}\right)^{1/2}$$

The most favorable region for overheating is the peripheral part of the plasma column, where there are magnetic islands with closed lines of force. A disruption of the local equilibrium in the tubes should cause them to "explode" and should cause a flux of energy into the relatively cool surrounding plasma as a result of collisional diffusion. In particular, this mechanism has been used to explain experiments in the TM-3. It should be recognized, however, that transport across the magnetic field in the plasma of a turbulent discharge has by no means received adequate study, either experimental or theoretical. One of the main reasons for this situation seems to be that experiments have not been carried out in large toroidal systems.

4 TURBULENT CURRENT HEATING OF PLASMAS IN LARGE CLOSED MAGNETIC CONFINEMENT SYSTEMS

The use of turbulent current heating in small closed magnetic confinement systems has met with definite success: Quite high temperatures have been achieved in times on the order of 1 or 20 μs; the current energy has been dissipated significantly faster than in the case of ordinary ohmic heating; and the large anomalous resistance has made it possible to achieve rapid penetration of the current into the plasma column.

A question which naturally arises is whether turbulent current heating can be used in large fusion devices, e.g., in a large-scale tokamak, to achieve the ignition temperature. This question is quite pertinent for two reasons. First, we know that classical ohmic heating will not get us to fusion temperatures, so that auxiliary heating will have to be used in a tokamak reactor. At present it would be hazardous to assert that the new methods which have been developed (neutral injection and rf heating) will not drive any strong instabilities in large-scale devices and thus will not cause a catastrophic increase in the energy loss.

Second, if the working point on the collision-rate scale for a tokamak reactor falls in a region of trapped-particle instabilities, then the plasma heating time must satisfy $\tau_{heat} < \tau_E \simeq a^2/\varkappa_{max}$. This requirement means that the energy must be pumped into the plasma very rapidly, and this can be achieved only by making use of the collective properties of the plasma. It thus seems completely reasonable to pursue the possibility of using turbulent current heating to reach the ignition temperature in a large toroidal device.[256] The final resolution of this question would have to come from a full-scale experiment, of course, but even at this point, with the data already obtained from small devices and with the present theory, it is possible to make a completely reasonable prediction, which will at least tell us whether a corresponding experiment would be worthwhile.

According to the experiments which have been carried out in small devices, the value of β_φ in a high-current discharge with a skin effect is established at the level $\beta_\varphi \approx 0.2\text{-}1$, where the value $\beta_\varphi \approx 1$ corresponds to discharges with high plasma densities.[83] It has also been shown that two types of turbulent discharges correspond to values $\beta_\varphi \sim 1$. In the first case, β_φ reaches values $\beta_\varphi \cong 1$ during the heating and is held there during the afterglow, while both the current and the thermal energy decay over time.[242] In the second case, the loss in the heating stage becomes so rapid that β_φ is kept at a level $\beta_\varphi \cong 1$, although energy continues to be pumped effectively into the plasma through dissipation of the current energy across the anomalous resistance.[234]

Experiments in a large straight system[6] with a current skin effect have shown that there is an effective method for transporting thermal energy from the skin layer into the interior of the plasma column at a rate such that the condition $\beta_\varphi \cong 1$ is maintained in the skin layer.

In large toroidal systems such as tokamak fusion reactors the turbulent-heating current would obviously flow in a skin layer with a thickness much less than the radius of the plasma column, $\delta \ll a$. Consequently, the establishment of the experimental facts outlined above is very important for carrying out calculations on the plasma heating by a turbulent current discharge in a large toroidal system.

Following the calculations of Hirose et al.,[235] we note that the current in the skin layer,

$$I = enu \cdot 2\pi\alpha\delta \qquad (\delta \ll \alpha),$$

will generate a poloidal field $H_\varphi = \frac{2I}{a}$. Hence

$$\delta = \frac{\sqrt{2}c}{\omega_{pi}\frac{u}{v_S}\sqrt{\beta\varphi}}, \quad \frac{kT_e}{e} = \frac{1}{2\pi}\frac{\beta_\varphi I^2}{a^2 ne}.$$

We now assume that a tokamak current with a distribution $j_{OH} = j_0\,[1-(r^2/a^2)]$ and a safety factor $q = rH/RH_\varphi$. is produced in the torus before the turbulent heating is begun. In the limit $r \to 0$, q has a minimum value $q_0 = H/2\pi j_0 R$, while in the limit $r \to a$ we have $q \to 2q_0$. If q is not to fall below q_0 in the skin layer when the turbulent-heating current begins to flow, the current must not exceed the total ohmic-heating current:

$$I_m = \frac{a^2 H}{2\pi q_0 R} = I_{OH}.$$

In a large toroidal system the current increase should apparently be limited by the inductance, rather than by the resistance. Calculations were carried out under this assumption. A current pulse was assumed to rise in a time $t = t_m$ and to have a constant amplitude $I_m = I_{OH}$ for $t = t_m$. It was assumed that $u/v_S > 1$ is a necessary condition for the driving of the ion acoustic instability.

Let us examine the stage of the current rise. We assume that T_e in the skin layer is determined from the expression

$$\frac{3}{2} k \frac{dT_e}{dt} = (1-\gamma^*)eEu .$$

Here γ^* is the fraction of the energy which is transported out of the skin layer. For simplicity we now assume that the conditions $\gamma^* = \text{const}$ and $\alpha = u/v_S = \text{const}$ hold during the stage of the current increase. We also assume that the heating in a dense plasma should occur at $u \simeq v_S$. We then find

$$\frac{\partial j}{\partial t} = nE \cdot \frac{du}{dt} = \frac{ne^2}{3M_i} \cdot \alpha^2 (1-\gamma^*)E$$

and thus $E/\delta^2 = 4\pi\partial j/\partial t$. Hence

$$\delta = \frac{\sqrt{3}c}{\omega_{pi}\alpha(1-\gamma^*)^{1/2}} .$$

An energy transfer out of the skin layer ($0<\gamma^*<1$) is thus reflected by an expansion of this layer. According to the expression derived previously for δ we have

$$\beta_\varphi = \frac{2}{3}(1-\gamma^*), \quad \frac{kT_e}{e} = \frac{1}{3\pi} \cdot \frac{(1-\gamma^*)I^2}{\alpha^2 ne} .$$

We can now determine the energy and the resistivity:

$$E = \frac{2\pi}{a} \delta \frac{dI}{dt}, \quad \eta = \frac{E}{neu} = 4\pi \frac{\delta^2}{I} \cdot \frac{dI}{dt}$$

At a constant value of δ the plasma resistance falls off as the current increases. This model obviously does not apply during the initial time interval in which the current density is increasing. If v_S is to follow u, we need a finite value of $j^2\eta$, and in the limit $j \to 0$ we need $\eta \to \infty$. Actually, the resistance is large but finite, and a correct calculation of the resistance would require incorporating the penetration of j and E into the plasma. The time required for penetration to a depth δ is short. If it is assumed that the noise level is $\frac{W}{nT_e} \approx 10^{-3}$ (this corresponds to the values observed in small toroidal systems), then we would have $\eta \sim \omega_{pe} \cdot \frac{W}{nT_e}$ according to the quasilinear theory. The penetration time $t_{\text{pen}} \sim \delta^2/\eta$ thus turns out to be on the order of 10^{-5} s for an assumed skin thickness of about 10 cm.

The total energy pumped in during the current rise time is given by

$$W_m = \frac{\sqrt{3}}{2} \cdot 4\pi c \cdot \frac{R}{a} \cdot \frac{I^2{}_m}{\omega_{pi} \cdot \alpha(1-\gamma^*)^{1/2}} .$$

This energy must be compared with the total energy of the plasma at the temperature required for ignition, T_{ig}:

$$W_{ig} = 6\pi^2 R^2 a^2 \bar{n} T_{ig}$$

There is an important point here: If the turbulent heating is carried out over a time much shorter than the time required for a temperature relaxation between the ions and electrons, then the electron temperature should be raised to $T_e \gtrsim T_{ig}$ by pumping in an energy $W_m \gtrsim W_{ig}$.

Estimates carried out for the large FPP and JET devices[256] have shown that the necessary temperatures in the skin layer, $(T_e)m$, can be reached, but the input energy level turns out to be lower than that required for ignition by a factor of 3-8. It thus becomes necessary to determine the amount of energy supplied during the subsequent stage of a constant current.

If the current is held constant, then an increase in the skin depth over time implies a decrease in u to values at which the condition for the occurrence of the ion acoustic instability is violated ($u > v_S$). It is natural to assume that the current heating will become ineffective under such conditions.

In the stage with $u/v_S > 1$ the skin depth changes in accordance with

$$\delta^2 = \frac{\eta}{4\pi}(t - t_m).$$

If we assume that β_φ and T_e in the skin layer are independent of the time and that we have η = const in the stage in which the condition $u > v_S$ holds, then we have $d\delta/dt = \eta/8\pi\delta$. The heating rate is

$$\frac{dW}{dt} = I^2 R_{pl} = \frac{R\eta}{a\delta} I^2 .$$

We then have $dW/d\delta = 8\pi \dfrac{R}{a} I^2$ and $W - W_m = 8\pi \dfrac{R}{a} I^2(\delta-\delta_m)$.

The auxiliary heating is thus determined by the increase in the skin depth. If we hold β_φ and v_S constant, then we have $u/v_S \sim \delta^{-1}$. Calculations for the FPP and JET show that during the constant-current stage it is possible to pump in additional energy W sufficient to satisfy $W \gtrsim W_{ig}$. It should be noted that this calculation model ignores the energy loss from the plasma during the heating pulse. Furthermore, the calculations are based on a transfer of heat from the skin layer into the interior of the plasma column, and this transfer must be checked in actual large-scale devices.

There are several questions associated with achieving MHD stability of the plasma column and of the skin layer during turbulent heating in a large toroidal system which require serious study. Nevertheless, qualitative estimates of the possibilities of turbulent heating in large toroidal devices are quite encouraging, and testing in a large experimental device is definitely warranted.

CHAPTER VI
MICROINSTABILITIES IN A TURBULENT CURRENT-CARRYING PLASMA

1 DEVELOPMENT OF THE PHYSICAL CONCEPTS AND METHODS FOR STUDYING COLLECTIVE INTERACTIONS IN PLASMAS

In the phenomenological picture of turbulent heating which was drawn above, and which gives a general description of the effective pumping of energy into a plasma for a particular loss mechanism, only a single condition was postulated: a stabilization of the drift velocity.

The appearance of a resistance in a collisionless plasma is indirect evidence of an effective scattering of electrons by plasma noise. Stabilization of the drift velocity apparently requires reaching the threshold for an instability which causes energy from the external source to be transferred through the ordered motion of electrons into waves of some type or other in the plasma. For a better understanding of the physics of turbulent heating it is thus extremely important to study the nature of the microinstabilities, to determine the frequency and wave-vector spectra, and to determine the level of the turbulent fluctuations excited in the plasma by the current.

It is equally important to study the evolution of the various types of turbulent spectra and to determine the nature and strength of their relationships. Information of this type can answer several questions associated with the loss mechanisms and the problem of the collective acceleration of particles in a turbulent plasma.

In order to reach an understanding of the highly turbulent state of a plasma it is necessary to study not only the average values of the plasma parameters (the temperature and density) over the cross section but also the local values. This need for local values substantially complicates efforts to comprehensively analyze the turbulence characteristics and to compare the results obtained in different experiments. Since the experimental information available on the turbulent noise level and on the spatial localization of various wave processes is incomplete, as is that on the wave conversion coefficients, it is extremely difficult at present to evaluate the role played by the superthermal emission in the plasma energy balance. It thus becomes equally important to study the mechanisms for the accumulation of turbulent noise and the extent to which this noise is localized in a plasma.

The first information on the plasma state in the case of strong turbulence came from astrophysics. The intense radar noise observed on February 27 and 28, 1942, by Hey (in England), using radar sets working at wavelengths of 4 and 6 m, constituted the first observation of superthermal radio emission from the active sun. Later, in 1947, Payne-Scott,[260] Wild,[303] and their colleagues studied type II radio bursts in the frequency range 50-200 MHz during solar chromospheric flares. In the dynamic radio spectra they detected narrow bands, a few tens of megahertz wide, corresponding to two emission peaks, whose frequencies decreased over time. The identical shapes of the emission lines and the characteristic intensity ratio of these lines constituted evidence that two plasma harmonics were being observed.

Shklovskii[181] (USSR) and Martin[252] (England) suggested that the solar radio bursts were associated with waves in a plasma. This hypothesis was subsequently extended in theoretical papers by Zheleznyakov and Ginzburg.[85,62] Collective effects in a plasma were studied by Akhiezer, Prokhoda, and Sitenko.[14]

The theory worked out in these papers showed for the first time that density inhomogeneities in a plasma give rise to the emission of waves. The efficiency at which a longitudinal plasma wave is converted into a transverse wave becomes particularly high in a region where the geometric-optics approximation is violated simultaneously for the longitudinal and transverse waves — where the dielectric function of the plasma is close to zero. When plasma waves encounter a large density gradient at small angles of incidence, about 10% of the plasma-wave energy flux can be converted into electromagnetic emission. Large-amplitude, steady-state plasma waves may be very nonsinusoidal, so that harmonics in the nonlinear wave could be responsible for the emission at harmonics of the plasma frequency. There is, however, another possible explanation for the radio emission near the frequency $2\omega_{pe}$, as Ginzburg and Zheleznyakov pointed out.[62] According to this hypothesis, a nonlinear coalescence of plasma waves would cause emission as the result of the Raman scattering of the "fundamental plasma frequency."

Calculations in Refs. 86 and 62 for the plasma

parameters of the solar corona ($n \approx 10^8$ cm^{-3}, $T_e \approx 2 \cdot 10^6$ K, and a conversion length on the order of 10^{10} cm) yielded a coefficient of about 10^{-5} for the conversion of longitudinal waves into transverse waves near the frequency ω_{pe}, and they yielded a coefficient of about 10^{-6} at twice the plasma frequency, $2\omega_{pe}$. The degree of turbulence of the solar plasma is $\frac{W_l}{nT} \approx 10^{-5}$.

Experimental confirmation of the theory worked out in those papers comes from the qualitative agreement between the calculated spectral intensity of the superthermal solar radio bursts and the intensity observed in the studies by Payne-Scott and Wild.[303]

The dynamic spectra of the sporadic solar radio bursts reveal a variety of typical forms, and the bursts are accordingly classified on the basis of duration, type of spectrogram, and intensity. Wild's scheme[303] for classifying solar bursts in the meter wavelength range, which puts all bursts in five basic categories, is now generally accepted. The fine structure of the sporadic solar radio bursts is presently the object of close theoretical and experimental study in astrophysics.

The sporadic radio emission from the active sun was not the sole space-physics observation of collective manifestations of a plasma on the eve of the research on turbulent current heating.

In 1958, Bowles[199] first observed a backscatter of a high-power 41-MHz radar pulse from the ionospheric F layer near the geomagnetic equator. In the same year, Gordon[219] studied ionospheric backscatter over a broader frequency range, from 50 to 100 MHz. The screen of the radar set revealed a long train of bursts, of decreasing amplitude, corresponding to the scatter from the ionosphere of a short radar pulse which had been emitted at the transmitter frequency. The signals scattered from the ionosphere agreed in order of magnitude with calculations based on the Thomson theory for the scattering by individual electrons, but no Doppler broadening, $\Delta\omega \sim \left(\frac{T_e}{m_e\lambda^2}\right)^{1/2}$, was observed, even though the apparatus was capable of detecting it.

The mechanism for the scattering of an intense radio wave under conditions of complete transparency of the ionospheric F layer ($\omega \gg \omega_{pe}$) was explained by Dougherty and Farlay[207] and Salpeter.[275] They showed that when the sounding wavelength λ_0 was equal to or greater than the average distance between the electrons in a collisionless plasma, i.e., under the condition

$$\lambda_0\left(4\pi\lambda_D \sin \frac{\vartheta}{2} \right)^{-1} \gtrsim 1 \qquad \text{(VI.1)}$$

it becomes necessary to consider collective effects associated with fluctuations of the refractive index of the medium. In this case the line broadening of the scattered radiation does not depend on the thermal fluctuations of the electrons, and it does correspond roughly to the ion velocity. In other words, it is determined by the sound velocity: $\Delta\omega = \pm k_s v_s$. Condition (VI.1) is now known as the Salpeter condition; here $\lambda_D = v_{Te}/\omega_{pe}$ is the Debye wavelength, ϑ is the scattering angle, and $\mathbf{k}_S$ is the wave vector of the acoustic wave.

In the spectrum of the radio signal scattered from the

ionosphere, Gordon[219] thus made the first observation of two peaks (satellites) at the characteristic ion plasma frequencies, shifted from the frequency of the incident signal.

Methods based on detecting the radio emission by the plasma itself and detecting the scattering of external electromagnetic waves in the plasma are the basic tools used to study collective plasma processes in space. For laboratory plasmas, in contrast, these methods are of limited applicability, and, as in the astrophysical applications, attempts to interpret the results run into many difficulties. In laboratory experiments on highly turbulent plasmas, however, these methods have been supplemented to a large extent and also developed, both theoretically and experimentally, along with other diagnostic methods, e.g., optical methods.

Research on microinstabilities in laboratory plasmas was basically stimulated by experiments on interactions of electron beams with plasmas. Classic theoretical and experimental studies of the interactions of beams with plasmas were carried out over the years 1948-1960 by Akhiezer and Faĭnberg,[12] Bohm and Gross,[198] and Faĭnberg.[166] Their laboratory experiments proved that the high amplitudes of the waves excited by an electron beam, over broad frequency and wave-number spectra, were responsible for the highly nonlinear phenomena and the effective heating of the plasma by the beam.

Bursts of decimeter-range electromagnetic radiation were first identified as superthermal radio emission near the plasma frequency, $\omega_{pe}/2\pi = 2\cdot 10^9$ Hz, in beam–plasma experiments by Demirkhanov, Gevorkov, and Popov[77] and by

Kharchenko and Faǐnberg.[173] These radio bursts were studied in detail by Nezlin,[139] Smullin and Gatti,[282] and Berezin.[34]

Kharchenko[174] and Berezin[34] experimentally studied the regularity of the waves excited in a beam by a plasma with a density $n = 10^{12}$-10^{13} cm^{-3}. The plasma noise was so intense that it could easily be detected by a broad-band oscilloscope with a traveling-wave deflection system over a frequency band up to 3000 MHz. Regular trains were observed for 4-10 ns; their duration depended on the plasma density and the beam current.

The physical groundwork for future experiments on turbulent current heating was laid in the 1960s. Budker[42] and later Buneman[201] examined the theoretical possibility of driving an instability of the beam type during the ordered motion of all the electrons with respect to the ions in a strong electric field, stronger than the Dreicer field.[208]

For a plasma in a longitudinal electric field, the initial conditions are fundamentally different from those for experiments in which a monoenergetic beam is injected into a plasma, and all the plasma particles should be involved in a collective acceleration. These arguments led to the concept of a plasma betatron – a concept which, as it turns out, was not to be implemented.

Skarsgard[279,232] and Stefanovskiĭ[157] showed experimentally that only a small fraction of the electrons are captured into free acceleration; most are effectively decelerated, dropping out of the acceleration in the very first stage of the process. The primary reason for the lack of acceleration was the onset of a small-scale instability.

In separate series of measurements in the Model B1 stellarator in electric fields $E \sim E_{Dr}$ in 1956-1958, Heald,[230] Bernstein,[192,193] and Coor[203] were among the first to observe intense superthermal microwave emission in plasma experiments. The excitation of longitudinal plasma waves could be ascribed to the appearance of a small fraction of runaway electrons. The spectrum was analyzed with a set of detector waveguide sections near 9, 35, and 70 GHz and also by a narrow-band superheterodyne receiver at 35 GHz. The observed microwave emission was limited to the frequency interval 35±2 GHz, which corresponded to the electron plasma frequency at a particle density $n \approx 1.5 \cdot 10^{13}$ cm^{-3}. At an energy of about 100 keV of the hard x-rays caused by the runaway electrons, the power level of the microwave emission reached about 0.1 W. These investigators were the first to call attention to the pulsed nature of the superthermal microwave noise of plasmas and the importance of this effect in explaining the sporadic solar radio emission. Much importance was attached to experiments on the microwave noise spectra in the large plasma devices existing at the time (AL'FA in the USSR and ZETA in England).

Highly sensitive microwave radiometers used by Harding et al.[172] and Larionov[126-130] revealed thermal radio noise and thus made it possible to measure the temperature of the plasma electrons. Under certain conditions in the AL'FA, however, at low densities, at which electron-runaway conditions occurred, Larionov[127,59] observed a sporadic superthermal microwave emission at a wavelength of 32 mm. The intensity of this emission was 100 times or more higher than that of the thermal microwave noise ($T_e \approx$ 100 eV).

Harding[172] and Wort[304] suggested that it might be possible to observe a nonthermal radio emission in the ZETA device. They thought that this emission might result from an effective scattering of plasma noise by plasma density fluctuations (see also Ref. 268).

There was thus mounting evidence that collective effects played a fundamental role and that it would be extremely difficult to produce a hot plasma without instabilities under laboratory conditions.

A test of the physical principles and mechanisms for turbulent heating required direct observation of instabilities in a current-carrying plasma and detection of intense turbulent noise. Turbulence in a plasma in a strong electric field was first detected by the present authors[159] in a straight discharge in the GROM-2 device; the turbulence was inferred from superthermal microwave emission. It was suggested (correctly) that the intense microwave emission near ω_{pe} ($\lambda \sim 8$ mm) and at lower frequencies resulted from a conversion of longitudinal plasma waves into electromagnetic waves and thus gave only a qualitative measure of the turbulent noise level in the plasma. Nevertheless, reliable detection of the superthermal microwave emission was the simplest and most reliable method for detecting the onset of a current-driven instability.

Convincing proof that the collective excitation of waves in plasmas by directed streams of particles with currents above a certain critical value is a general physical process came from experiments in 1964-1966 in some very different devices.

Demidov, Fanchenko, et al.[73,72,167] observed superthermal microwave emission and x-ray emission in a toroidal electrodeless discharge in strong electric fields directed parallel to the confining magnetic field, under conditions with an anomalous plasma resistance. In these particular experiments, the Raman scattering of an intense electromagnetic signal with a wavelength $\lambda_0 = 3$ cm by plasma waves was observed in a plasma with $n = (2\text{-}5) \cdot 10^{11}$ cm^{-3}, $T_e = 10^2\text{-}10^3$ eV, $E = 50$ V/cm, and $H = 4$ kOe. It was noted that the plasma emission was much fainter in the long-wave part of the spectrum ($\lambda \approx$ 1.9-22 cm) and that the emission occurred only near the "violet" satellite ($\lambda \approx$ 1.7-2.1 cm) when a turbulent plasma was bombarded with a microwave signal with a pulsed power of about $(1\text{-}5) \cdot 10^3$ W. For the first time, therefore, plasma experiments had revealed a high level of plasma waves, with a degree of turbulence amounting to $\frac{W_l}{nT} \sim 10^{-1}$ and with a typical wave number $k \sim \frac{k_D}{6}$.

A turbulent plasma with a high plasma-wave noise level was also produced at the front of a collisionless cylindrical shock wave (Refs. 21 and 125).

Adamov et al.[1] observed superthermal microwave emission from an induction θ pinch with preionization at the times at which the critical plasma density was reached for the corresponding wavelength of the detector section, in the wavelength range $\lambda \approx$ 4-32 mm. A correlation between the microwave emission and the x-ray emission was observed. A betatron mechanism for the acceleration of electrons across the magnetic field was proposed in an effort to explain the onset of turbulence in the plasma.

The turbulent state and the spectrum of the superthermal sporadic microwave emission near ω_{pe} and $2\omega_{pe}$ were studied during the annihilation of oppositely penetrating magnetic fields in a fully ionized hydrogen plasma by one of the present authors.[141,134] For the first time in a dense plasma, with $n \approx 6\cdot10^{13}$ cm^{-3} and $T_e \approx$ 65 eV, the Raman scattering of a wave with λ_0 = 4 mm revealed a high plasma-wave level, $W_l \sim 10^{-3}nT$. These waves were excited against a higher background of ion-plasma-wave noise, ω_{pi}, and could not have been a consequence of classical electron runaway, as in the case (for example) of strong electric fields parallel to the magnetic field.

A high plasma-wave level was inferred from the Raman scattering of a signal at λ_0 = 3 cm in experiments on a turbulent straight discharge by Kalinin et al.[112] The noise level reached ($W_l \sim 0.1$-$0.2nT$), and the typical turbulence wave vector was $\mathbf{k} \sim \mathbf{k}_D$.

In all of the highly turbulent discharges, we might note, the plasma-wave level was high enough to lead to noise emission near the second harmonic of the plasma frequency, $2\omega_{pe}$ (Refs. 73, 141, and 112). The ratio of the emission intensities at the first and second harmonics in these experiments was typically $P_{\omega_{pe}} \sim (10$-$10^2)P_{2\omega_{pe}}$.

Finally, experiments in approximately steady-state toroidal discharges — in stellarators and tokamaks — were an extremely important step in research on collective processes in hot plasmas. We have already mentioned the first observations of sporadic microwave emission and hard x-ray emission in the Model Bl stellarator.[230,192] More-detailed information was obtained by Blinov in the S-l stellarator.[37] These measurements were carried out in a helium plasma over broad ranges of the electric and magnetic fields (E = 0.05-0.5 V/cm, H = 2.5-10 kOe) at a density $n \sim 10^{12}$-10^{13} cm^{-3}. The basic conclusion of that study was that a significant fraction of the electrons (about 10^{-3} of the total number carrying the current) are accelerated in a racetrack stellarator, and at a certain threshold number of accelerated particles an instability occurs, accompanied by intense radio emission near the electron plasma frequency. An interesting fact established in the S-1 experiments was that the threshold for the excitation of sporadic bursts in the microwave range depended on the strengths of the electric and magnetic fields.

In experiments under anomalous-resistance conditions in the SIRIUS stellarator,[46] in strong electric fields, $E \sim (1-10)E_{Dr}$, at which the plasma resistance increased from 1 to 8 Ω, intense microwave emission was observed over the wavelength interval λ = 2-4 cm, i.e., near the plasma frequency. In weaker electric fields, $E < E_{Dr}$, the maximum in the emission spectrum shifted toward longer wavelengths, λ = 12-15 cm, which corresponded to the Buneman frequency [Eq. (III.3)]. The peak emission power from the entire surface area of the plasma column in the wavelength interval $\lambda \approx$ 2.6-3.6 cm was about 100 W, while that in the interval $\lambda \approx$ 7.5-10 cm was about 1000 W.

Superthermal sporadic emission in the T-5 and TM-1 tokamaks was studied by Gorokhov and Dolgov-Savel'ev.[67,68] The emission spectrum was analyzed with diffraction filters; the detector was a cooled indium antimonide cell (a Putley detector[267]). Microwave bursts lasting on the order of

100 μs with a radiance of about 10^3 W/(sr·cm^2) were observed. The spectrum from a tokamak discharge with $n \approx 10^{13}$ cm^{-3}, H = 10 kOe, and T_e = 40 eV exhibited a sharp decay from the plasma frequency to the upper hybrid frequency ω_h ($\frac{\omega_{pe}}{2\pi} \approx 2.8 \cdot 10^{10}$ Hz, $\frac{\omega_h}{2\pi} \approx 3.8 \cdot 10^{10}$ Hz). The emission frequency was observed to depend on the plasma density; this dependence might have been responsible for the shift of the peak in the emission spectrum during a single discharge.

The early experiments carried out to detect and study the highly turbulent plasma state established the following basic experimental facts.

1. There is a clear temporal correlation between the occurrence of the anomalous resistance and the onset of plasma-wave turbulence [$W_l \sim (10^{-1}\text{-}10^{-3})nT$], and the anomalous resistance is accompanied by a high level of superthermal microwave emission near the plasma frequency ω_{pe}.

2. The differences in the plasma turbulence spectra are more probably the consequence of different discharge regimes than of the differences among the devices.

3. As a rule, an interaction of a beam nature in the plasma and the highest level of superthermal microwave emission, correlated in time with the emission of hard x radiation, are observed in weak electric fields, $E \lesssim E_{Dr}$, in approximately steady discharges and in strong electric fields in pulsed discharges, with low densities and low temperatures ("startup" emission).

4. Under conditions of strong turbulence (in strong electric fields, $E > E_{Dr}$), a broader noise spectrum is

excited, $\omega_{pi} \lesssim \omega \lesssim \omega_{pe}$, and there is no correlation between the hard x-ray emission and the microwave emission.

The problem of identifying the instabilities thus required a further development of the theory of turbulent plasmas.

2 PRESENT UNDERSTANDING OF THE DYNAMICS OF TURBULENT PLASMAS

The basic problems involved in the study of plasma turbulence were first formulated by Kadomtsev.[107,109] The theory of collective interactions in plasmas developed extremely rapidly in 1965-1970. The basic types of streaming instabilities were studied, and the states of strong and weak plasma turbulence were classified in terms of not only absolute noise energy density but also how well the high-frequency and low-frequency spectra are coupled. A generalized theory of plasma turbulence based on a kinetic model for the collective excitations of so-called plasmons, incorporating decay processes and induced scattering by waves and particles, was derived by Romanov and Filippov[145]; Vedenov, Velikhov, and Sagdeev[55]; and Drummond and Pines.[209] This theory was subsequently put in systematic form by Rudakov,[147,149] Tsytovich,[176] and Galeev and Sagdeev.[60]

A major point in the theoretical generalizations of the collective interactions in plasmas, formulated by analogy with the turbulence of an incompressible fluid, is that the energy of the turbulent fluctuations is not lost as it is pumped over the wave-number spectrum. The spectrum can be divided into regions in which fluctuations are excited and regions in which they are absorbed. The operation of effective energy-conversion mechanisms in a plasma causes energy to be removed from regions in which waves are excited and transferred to absorption regions. The result is a heating of the plasma. On the other hand, these conversion mechanisms create conditions for the emission of waves and for the acceleration of charged particles. The transfer of wave energy is directed from larger to smaller wave numbers, because of the nonlinear interactions of the plasmons with each other and of plasmons with particles.

For plasma waves, according to Tsytovich,[175] the induced scattering by electrons goes quite rapidly at $v_\varphi \ll 10 v_{Te}$:

$$\frac{1}{\tau_l} \simeq \left(\frac{v_{Te}}{v_\varphi} \right)^3 \frac{W_l}{nT} \omega_{pe}.$$

If the energy loss in such conversions in the plasma is small, energy builds up to a value $W_l \approx \left(\frac{k}{k_D} \right)^2 nT$ at small wave numbers, even when the turbulence source is relatively weak. A so-called plasmon condensation occurs and can give rise to a modulational instability.[111]

The possible instability of a cold gas of Langmuir plasmons with respect to self-contraction was first pointed out by Vedenov and Rudakov.[56]

Galeev and Sagdeev[60] showed that a modulational instability results from the rf pressure exerted by the waves trapped by a slight density fluctuation and from the subsequent displacement of plasma out of a bounded region by the electromagnetic wave.*

Zakharov[99] examined the nonlinear stage of a Langmuir condensate, in which, upon the formation of a rarefaction (a caviton), density fluctuations grow to a level $\frac{\delta n}{n} \simeq 1$, causing a collapse. According to Khakimov and Tsytovich,[171] the onset of a modulational instability causes wave energy to be pumped from slow fluctuations to fast ones. The largest wave vector $\mathbf{k}_M$ in the modulational-perturbation spectrum is on the order of $\mathbf{k}_M^{max} \simeq \mathbf{k}_D \frac{We}{nT}$, i.e., in the case of strong Langmuir turbulence, the caviton velocities reach the ion acoustic velocity, on the average:

$$\frac{\omega}{\mathbf{k}_M} \simeq v_{Te} \left(\frac{W_l}{nT}\right)^{1/2} \sim v_{Te} \sqrt{\frac{m_e}{M_i}}.$$

As was shown by Degtyarev and Zakharov,[71] Shapiro et al.,[180] Rudakov,[149,270] Khakimov and Tsytovich,[171] and Weinstock and Bezzerides,[299] a basic aspect of the strong turbulence which results from the modulational instability is that the wave condensate may be essentially destroyed because of an inverse energy transfer from small wave numbers to large wave numbers. Part of the condensate energy may be expended on the emission of electromagnetic waves and the acceleration of particles.

The derivation of a theory for the modulational instability was an important step in research on the dynamics of plasma turbulence. This theory put a new light on several effects which had been observed in a

---

*See also: T. F. Volkov, "Effect of an rf electromagnetic field on plasma waves," in: Plasma Physics and the Problem of Controlled Thermonuclear Reactions, Vol. 4, Pergamon, New York (1961).

highly turbulent plasma, for example, the sporadic superthermal microwave emission, the interaction of high-frequency and low-frequency spectra, and the formation of the tails on the electron and ion distribution functions.

Effects associated with strong wave interactions in plasmas were tested in several experiments.

Kim, Stenzel, and Wong[244] observed a caviton (a localized region) in an inhomogeneous plasma at the Langmuir-resonance point for the given frequency of the electromagnetic pump wave ($\frac{\omega_0}{2\pi} = 9.6 \cdot 10^9$ Hz). In the absence of a magnetic field, the wave displaced the plasma and created a local cavity, from which accelerated particles were observed to stream in the direction perpendicular to the incident electromagnetic pump. Stenzel[284] detected a self-contraction of cavitons which were stretched out along the magnetic field during high-frequency pumping by large-amplitude whistlers in a krypton plasma of large dimensions ($0.05 < \frac{\omega}{\omega_{ce}} < 1$, $n = 10^{12}$ cm^{-3}, $T_e \simeq 2$ eV $\simeq 10T_i$, $\frac{\omega_{pe}}{\omega_{ce}} \gtrsim 10$, $l = 300$ cm, $a_{pl} = 45$ cm). Antipov, Nezlin, et al.[4] observed a modulational instability of plasma waves excited by an electron beam in a collisionless plasma with $n = 3 \cdot 10^9$ cm^{-3} and $T_e \approx 10$ eV in a magnetic field $H = 2$ kOe. Wave bunches appeared, created density wells, and localized in them, forming steady-state Langmuir solitons. The soliton lifetime amounted to tens of thousands of times the periods of the plasma waves.

The onset of a current-driven turbulence in a strong electric field directed parallel to a magnetic field is one of the most complicated effects. The plasma, a system with many collective degrees of freedom, can go through a series of states during the onset of the turbulence. In these states, important changes occur in the plasma parameters and, correspondingly, in the wave spectra themselves. Furthermore, several coupled wave branches can arise simutaneously in a magnetized plasma. These circumstances are sources of difficulties in attempts to compare theory with experiment.

As Rudakov and Korablev[147] and also Stringer[286] have shown, the ion acoustic turbulence works through an effective friction to oppose the particle acceleration caused by the electric field in a nonisothermal plasma ($T_e >> T_i$) and thereby causes the anomalous-resistance effect.

In the case of weak turbulence the ion sound has a nondecay spectrum, and three-plasmon wave processes cannot damp it (Kadomtsev and Petviashvili[107]). Accordingly, induced scattering by ions should play a leading role in shaping the steady-state acoustic spectrum. The coupling of these waves with electrons is weak because of the blurring of the resonance in the wave–particle interaction (Rudakov and Tsytovich[269]). As a result, energy is transported in wave-number space over a characteristic time[143,109] $\frac{1}{\tau_S} \sim \omega_{pi} \frac{W_S}{nT_i} \vartheta^2$, where ϑ is the angle between the propagation directions of the interacting waves.

The energy spectrum of ion acoustic turbulence was first derived by Kadomtsev and Petviashvili[107]:

$$W_S \sim \frac{1}{k^3} \ln \frac{\mathbf{k}_D}{\mathbf{k}}$$

The noise energy turns out to be large at small wave numbers $\mathbf{k}$ although (as mentioned earlier) the pumping due to the instability is most intense near the upper boundary of the wave-number range, $k \sim k_D$.

A magnetic field introduces an additional mechanism for strong absorption of ion acoustic waves, by ions near the ion cyclotron resonance, $k \approx \frac{\omega_{ci}}{v_S}$; on the other hand, a magnetic field intensifies the conversion of the sound into "whistlers," i.e., weakly damped electromagnetic waves which propagate primarily along the magnetic field. This conversion into whistlers, which is important for the diagnostics of turbulent plasmas, was pointed out by Rudakov.[113] Whistlers, which have frequencies

$$\omega_k^\Omega = \frac{k^2c^2|\omega_{ce}||\cos\vartheta|}{\omega_{pe}^2 + k^2c^2},$$

can be excited by an anisotropic instability as a result of the conversion of ion acoustic and plasma waves (Tsytovich[175]). The possible excitation of whistlers was first pointed out by Camac et al.[202]

In a nonisothermal plasma ($T_e \gg T_i$), in which conditions are set for the excitation of an ion acoustic turbulence, the overlap of the frequency spectra of the whistlers and of the ion acoustic waves at $\omega_{pi} > \omega_{ci}$ allows energy to be transferred from the ion acoustic waves to the whistlers. Consequently, the spectra of both the whistlers and the ion acoustic fluctuations depend to some extent on the interaction between these waves. According to Kingsep,[118] the induced scattering of ion acoustic waves by ions leads to the excitation of whistlers over a time

$$\frac{1}{\tau_\Omega} \approx \omega_k^\Omega \left(\frac{T_i}{T_e}\right)^2 \frac{W_S}{nM_i v_A^2}$$

if the waves are isotropic. Whistlers continue to be excited until the energy loss resulting from their scattering balances the energy transferred to them from ion acoustic waves or until the ion acoustic intensity falls off because of the excitation of whistlers. The result may be the establishment of steady-state spectra:

$$W_{\omega}^{s} = \frac{\text{const}}{\omega^2}, \quad W_{\omega,\vartheta'}^{\Omega} = \text{const}.$$

A second mechanism for whistler excitation may be a decay process in the ion acoustic spectrum, $\Omega \leftarrow S' + S''$, although decay processes are allowed in only a relatively narrow interval of angles ϑ' and of frequencies for the maximum wave vectors k_S'; $k_S'' \simeq \frac{\omega_{pe}}{v_{Te}} \simeq k_D$. The decay mechanism for the excitation of whistlers is analogous to a mechanism for the emission of plasma waves near the second harmonic of the plasma frequency (this other mechanism was studied previously by Ginzburg and Zheleznyakov,[62] Aamodt and Drummond,[183] and Ivanov and Ryutov[102]). In contrast with the electromagnetic waves at the sum frequency $2\omega_{pe}$ which are emitted by the plasma, however, the whistlers are natural, weakly damped electromagnetic waves which are capable of growing in a plasma. The growth of whistlers over time depends on the angle involved in the interaction of the ion acoustic waves and the energy density of these waves:

$$\frac{\partial W_{\omega,\vartheta'}^{\Omega}}{\partial t} = \frac{v_S \omega_{pi}^{3} \sqrt{\omega} \sin^2 \Theta}{c^3 |\cos \Theta| \sqrt{|\cos \Theta|} \, n M_i v_A^{2} \omega_{ci}^{3/2}} \left(\frac{W_{\omega}^{S}}{2 v_S} \right)^2.$$

The whistler frequency spectrum is near $\omega^{\Omega} \simeq 2\omega_{pi} \simeq \omega_S' + \omega_S''$, where the whistlers can easily be detected experimentally, since $k_{\Omega} \gg k_S'$; k_S'' . In other words, the wave numbers of these waves are relatively large.[113,142]

One of the typical types of low-frequency turbulence excited by a transverse current in a magnetized isothermal plasma ($T_e \simeq T_i$) and in a nonisothermal ion-hot plasma ($T_i > T_e$) is the electron acoustic turbulence discussed by Sizonenko and Stepanov.[151]

As mentioned earlier, the problem of identifying the various types of low-frequency turbulent noise experimentally is an excellent example of concentrated complexities. As a consequence, we still lack a really good understanding of the dynamics of the ion plasma turbulence, and the subsequent evolution of the low-frequency spectrum in a highly turbulent magnetized plasma remains an open question.

The mechanisms for the excitation of plasma waves in a plasma with a pronounced ion acoustic turbulence are no simpler to interpret, particularly for the plasma in a closed magnetic confinement system with a strong electric field $E > E_{Dr}$. Because of the electron runaway and the separation of particles into "trapped" and "passing" (or "untrapped"), the longitudinal-velocity distribution becomes greatly distorted. Regions with a positive derivative or plateaus may form at the tail of the particle distribution (Kadomtsev and Pogutse[241]).

A theory derived for the stability of such distributions by Parail and Pogutse[140] and Coppi et al.[204] shows that plasma waves can be excited with a spectrum $\omega = \frac{k_\parallel}{k}\omega_{pe}$ in a magnetized plasma by virtue of an anomalous Doppler effect near the cyclotron resonance, $\omega = k_\parallel v_\parallel - m|\omega_{ce}|$ ($m = 1, 2, 3, \ldots$). The elastic scattering of superthermal electrons by these waves can cause a conversion of longitudinal electron energy into transverse energy, with $T_e^{\perp} \sim T_e^{\parallel}$; the instability involving relativistic electrons is a relaxation instability.[140] Furthermore, the Cerenkov resonance $\omega = k_{\parallel} v_{\parallel}$ may lead to the excitation of a spectrum of ion plasma waves.[204]

Experiments in the large ALCATOR,[204,276] TFR,[293] and TM-3 (Ref. 140) tokamaks, with a stream of superthermal electrons, $v > v_{Te}$, have in fact resulted in the driving of kinetic instabilities with frequency spectra $\omega \sim \omega_{pi}$ and $\omega \lesssim \omega_{pe}$. These results are generally consistent with the physical arguments regarding the beam mechanism for the excitation of noise by runaway electrons.

As mentioned earlier (Section 1 of Chapter VI), however, the very first experiments on turbulent heating under anomalous-resistance conditions revealed that the intense superthermal microwave emission was not correlated in time with the emission of hard x rays with energies $(10\text{-}10^2) \cdot T_e$. Furthermore, the runaway effect does not occur in experiments on the annihilation of oppositely directed magnetic fields and on shock waves in plasmas, nor is runaway predicted by the theory of Ref. 60.

Tsytovich[294] has analyzed two possibilities for the generation of a Langmuir turbulence in the presence of ion sound for a plasma model for the solar corona. The first of these possibilities also involves the runaway of some of the electrons from the tail of the distribution, since the frictional force corresponding to turbulent collisions is analogous to a Coulomb force: It increases linearly at $v \lesssim v_{Te}$ and then falls off as v^{-2}. The bulk of the electrons, moving under the influence of the field E, have a velocity $v = u$ in accordance with the balance between the driving force eE and the frictional force. For the electrons with $v > v_r$, where $v_r = v_{Te}\sqrt{v_{Te}/u}$ the frictional force is always weaker than the driving force eE. As a result, a runaway occurs, triggers the classical beam-instability mechanism, and leads to the excitation of a Langmuir turbulence.

The second possibility involves the emission of plasma waves by ion acoustic waves in an induced bremsstrahlung process, as a result of effective collisions of fast electrons with ion acoustic waves. In the course of these collisions the electron momentum changes direction, and a recoil effect occurs in the form of a plasma wave. The growth rates for these two possible processes are of the same order of magnitude:

$$\Gamma = \frac{1}{5}\omega_{pe}\left(\frac{u}{v_{Te}}\right)\frac{W_S}{nT}.$$

A Langmuir turbulence can thus be excited over a time equal to some tens of ion-plasma-wave periods.

As various mechanisms for the conversion of longitudinal plasma waves come into play, electromagnetic radiation may be produced near the plasma frequency and its harmonics. For example, the power radiated at the plasma frequency from a unit volume of an unbounded, isotropic, turbulent plasma during the scattering of plasma waves by ion acoustic waves was found in Refs. 16, 183, 65, and 102:

$$P_{\omega_{pe}} \approx 10\,\omega_{pe}\left(\frac{\omega_{pe}}{ck_0}\right)\frac{W_l W_S}{m_e n c^2}. \qquad \text{(VI.2)}$$

Correspondingly, in an interaction of Langmuir plasmons there is emission at frequencies near $2\omega_{pe}$:

$$P_{2\omega_{pe}} \approx 30\,\omega_{pe}\left(\frac{\omega_{pe}}{ck_0}\right)^3\frac{W_l^2}{m_e n c^2}. \qquad \text{(VI.3)}$$

The scattering of external electromagnetic waves by plasma waves was also examined in these theoretical papers. This scattering can be described by the simple expression

$$P^{s}_{\omega_0+\omega_{pe}} \sim S_0 \left(\frac{\omega_0}{\omega_{pe}}\right)^2 \left(\frac{\omega_{pe}}{ck_0}\right)^3 \frac{\omega_{pe}}{c} \frac{W_l}{m_e c^2 n}. \qquad \text{(VI.4)}$$

The approximate results in (VI.2)-(VI.4) are important for the diagnostics of turbulent plasmas. Experimentally, the degree of plasma turbulence $[W/(nT)]$, the typical wave number k_0, and the relative degree of localization of the noise energy in the ion acoustic and Langmuir turbulence spectra (W_l/W_s) are determined from the emission spectra at the plasma-wave harmonics ω_{pe} and $2\omega_{pe}$ and also from the power of the scattered signal.

Further theoretical work on the diagnostics of turbulence spectra $W(k, \omega)$ through the use of the natural microwave emission from an anisotropic plasma was carried out by Ichimaru and Starr.[236] They showed that this emission is usually a factor of 10^4 more intense than the waves scattered by noise when external waves are incident on a plasma. An analysis by Lashmore-Davies[247] has shown that with a monochromatic pump, e.g., near the plasma frequency, with $k \to 0$, the threshold for the modulational instability is equivalent to that for the vibrational two-stream instability. An electromagnetic standing wave is excited in the latter pump wave, and density perturbations occur at the zero beats. This process can result in a conversion of plasma waves. Fidone et al.[215] have derived a theory for the superthermal microwave emission near $\omega \approx \omega_{pe}$.

It must be realized that our present understanding of the nature of turbulence is imperfect. Neither experiments nor the theory as it exists today can give us an unambiguous description of the events which occur in the highly turbulent state, i.e., under conditions such that the spectra and correlations of the turbulent fluctuations in a plasma are determined by not only the level of these fluctuations but also (and substantially) by their interactions. It is this state of turbulent motion of a plasma which arises during current heating in strong electric fields.

We turn now to the basic characteristics of the turbulent noise and of the microwave emission which have been observed in various experiments on turbulent heating.

3 *LANGMUIR TURBULENCE AND SUPERTHERMAL MICROWAVE EMISSION*

Recent advances in microwave spectral-analysis capabilities have revealed several new pieces of information about the nature of the superthermal microwave emission from a turbulent plasma.

An explosive, sporadic microwave emission was observed by the present authors[159] in the straight GROM-3 discharge with $n_e \simeq 10^{15}$ cm^{-3} and $T_e \simeq 1$ keV. At a time near the maximum of the discharge current, a short microwave burst occurred over a broad frequency band from ω_{pi} to ω_{pe} ($\lambda \sim 100$-0.8 mm). This burst was observed at the time at which the plasma resistance and diamagnetism increased and at which the spectral lines broadened (Fig. 9). This type of microwave noise is different from the "startup" noise which is generated at the beginning of a discharge at lower plasma parameters ($n \lesssim 10^{13}$ cm^{-3}, $T_e \approx 10$ eV, degree of ionization of 2%).

Startup microwave noise appeared near the plasma

frequencies ω_{pe} and $2\omega_{pe}$. The intensity of this noise was six orders of magnitude greater than that of the thermal noise. It was correlated in time with the signal indicating the current of fast electrons, measured with a Faraday cup positioned directly behind an aperture in the anode. The intensity of the startup emission depended slightly on the bank voltage, the working-gas pressure, and the magnetic field.

During the startup stage of the discharge in the GROM-3, in which the electrodes played a particularly important role as electron emitters, the nature of the collective interactions was thus similar to the events which occur when an external electron beam is injected into a plasma.

The beam–plasma mechanism for the excitation of superthermal microwave noise, however, may operate in the final stage of a discharge with electrodes as well as in the initial stage. In the experiments by Zavoĭskiĭ et al.[95] in a straight discharge, for example, the stage of turbulent current heating was followed by the driving of a current–convective instability, which cooled the plasma and produced fast electrons (see also the papers by Kalinin et al.[115,116]).

Figure 74 shows the general characteristics of the discharge in the NPR-2 device. The diamagnetic signal (*c*) begins increasing when the current starts to flow (*a*) in a preinjected hydrogen plasma; in 1.7 μs, the diamagnetic signal reaches $nT \simeq 2\cdot 10^{15}$ eV/cm^3. There is no microwave emission in this stage. Intense emission (*b*) arises after the nT signal decays (after 8 μs), as a result of a current–convective instability. The Langmuir turbulence

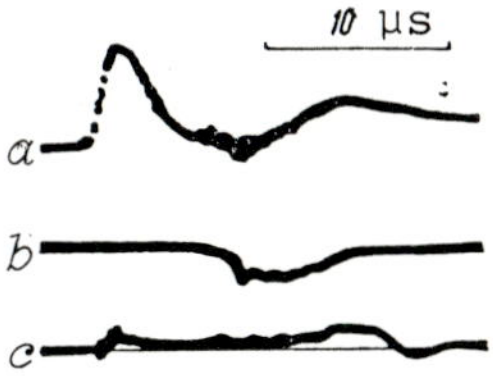

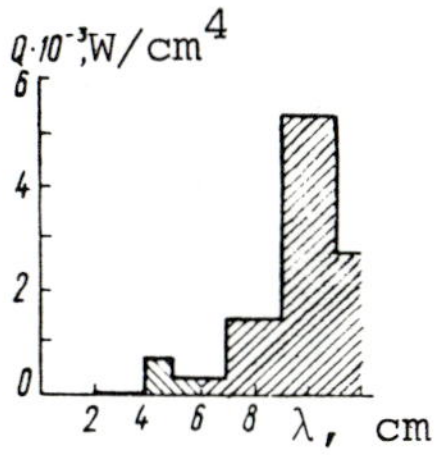

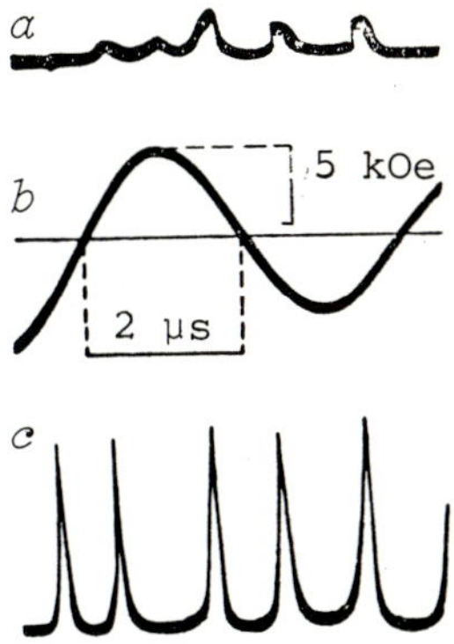

FIGURE 74 General characteristics of the turbulent discharge in the NPR-2 device.[95] *a*) Current in the plasma, $2 \cdot 10^{15}$ eV/cm^3 ; *b*) integrated electromagnetic emission from the plasma at $\lambda \sim 30$ mm; *c*) diamagnetic signal. $n \approx 10^{12}$ cm^{-3}, $H_0 = 2.5$ kOe, $E = 300$ V/cm.

FIGURE 75 Spectral power density of the emission from the plasma in the NPR-2 (Ref. 112). $n = 1.4 \cdot 10^{11}$ cm^{-3}, $T_e = 5 \cdot 10^2$ eV.

FIGURE 76 Oscilloscope traces illustrating the operation of the magnetic microwave spectrum analyzer of Ref. 95. *a*) Emission spectrum of the turbulent plasma in the NPR-2 in the wavelength interval $\lambda \sim 2$-4.6 cm; *b*) magnetic field amplitude in the analyzer; *c*) emission spectrum of a reference klystron source. $\lambda_0 = 3.4$ cm.

was studied by observing the Raman scattering of an electromagnetic signal with a wavelength $\lambda_0 = 3$ cm and also by detecting the superthermal microwave emission from the plasma (Kalinin et al.[112]). A tuned wavemeter was used to analyze the noise spectrum over the wavelength interval $\lambda = 2$-12 cm.

Figure 75 shows the spectral power density of the plasma-wave emission from the NPR-2 device. This spectrum has two lines, corresponding to the plasma frequency, ω_{pe} ($\lambda \simeq 10$ cm), and its second harmonic, $2\omega_{pe}$ ($\lambda \simeq 5$ cm). The ratio of the total amounts of power emitted at the first and second harmonics is found from the areas of the hatched regions in Fig. 75 to be $\frac{P_{2\omega_{pe}}}{P_{\omega_{pe}}} \cong 6\times10^{-2}$. According to Eqs. (VI.2) and (VI.3), this result corresponds to an equiprobable energy distribution among the degrees of freedom of the various turbulent waves in these experiments, W_l and W_i. The degree of turbulence and the characteristic wave vector of the waves in the plasma (n = $1.4\cdot10^{11}$ cm^{-3}, $T_e \approx 5\cdot10^2$ eV) were estimated from the data on the Raman scattering signal. The results were $\frac{W_l+W_s}{nT} \sim 10^{-2}$ and $k_0 \approx 9$ cm^{-1} $\ll k_D$.

Microwave spectral measurements were carried out for a denser plasma ($n \simeq 10^{12}$ cm^{-3}) in the NPR-2 during a single discharge with the help of some refined apparatus: a Zavoĭskiĭ analyzer, which is a magnetic analyzer which works by a paramagnetic-resonance effect.[89]

The emission spectrum in Fig. 76, obtained with a high time resolution, also reveals some localized peaks near ω_{pe} and $2\omega_{pe}$. The data on the noise intensity were used to estimate the energy densities of the plasma waves and the ion acoustic waves; the results turned out to agree well with data obtained previously in this apparatus.[112]

In high-current straight discharges under conditions such that there is a large number of fast electrons, the beam–plasma mechanisms are thus the most important and the most efficient mechanisms for the excitation of the extremely intense electromagnetic radiation near harmonics of the electron plasma frequency, ω_{pe} and $2\omega_{pe}$. According to a theoretical analysis by Zavoĭskiĭ et al.,[95] the growth rate of the plasma waves in a turbulent plasma upon the appearance of a beam of fast electrons is the net result of a superposition of three processes:

$$\delta\Gamma_l \sim \Gamma_l - \Gamma_S - \Gamma_{pi} \sim \omega_{pe} \left[\frac{n_e}{n} \frac{v_n^2}{(\Delta v)^2} - \frac{m_e}{M_i} \left(\frac{v_n}{v_{Te}} \right)^2 \frac{W_{i(S)}}{nT} - \frac{W_{i(pi)}}{nT} \right],$$

where Γ_l is the growth rate for the beam instability and corresponds to the first term, Γ_s corresponds to scattering of plasma waves by long acoustic waves (the second term), and Γ_{pi} corresponds to the induced scattering of plasma waves by electrons, accompanied by reemission in ion plasma waves (the third term). A Langmuir instability is driven if $\delta\Gamma_l > 0$, beyond a threshold value of the beam current, $\left(\frac{n_e}{n}\right)_{cr}$, and it is stopped by the scattering of plasma waves by electrons. The saturation level of the instability is

$$\frac{W_l}{nT} \approx \left(\frac{v_n}{v_{Te}} \right)^3 \delta\Gamma_l \omega_{pe}^{-1}. \tag{VI.5}$$

Knowing the degree of turbulence of the plasma and the temporal characteristics of the signals, we can use (VI.5) to determine the energy of the fast electrons, which are stabilized by the four processes listed above. In the NPR-2 device, the value of $\frac{v_n}{v_{Te}}$ was approximately 10.

Returning to the microwave emission from a highly turbulent, dense plasma, we may assert that the energy spectrum of the microwave emission in the GROM-3 device[288] showed a clear tendency for an increased emission intensity at long wavelengths (Fig. 9). The fact that the noise is excited only after a threshold in the plasma parameters is reached, on the one hand, and the spectrum, on the other, agree well with the theory of Ref. 147 for the ion acoustic instability. A somewhat unexpected result of these experiments was the emission near the electron plasma frequency,

$\lambda_{pe} \approx$ 2-0.8 mm, which implied the simultaneous excitation of a Langmuir turbulence and, possibly, a nonlinear mechanism for the conversion of plasma waves by turbulent fluctuations.

The first experimental test of the hypothesis of a nonlinear conversion of plasma waves in a turbulent plasma was carried out in the experiments by Zavoĭskiĭ et al.[92] in the VIKhR'-2 device.

Since the plasma column which forms in this geometry, with a very bumpy toroidal magnetic field, is highly inhomogeneous in the radial and azimuthal directions and is isolated from the chamber wall,[75,76,169,214] the emission spectrum was identified by varying the plasma density and by making a quantitative comparison of the wavelength of the short-wave part of the spectrum with the maximum density in the discharge. The spectrum was measured over the wavelength interval $\lambda \approx$ 40-2 mm with a two-channel system of detector receivers, whose output signals were fed to broadband oscilloscopes.[206] Figure 77 shows oscilloscope traces of the signals corresponding to the microwave emission from the turbulent plasma in the VIKhR'-2 device for two values of the initial plasma density. We see that over the entire frequency range studied the microwave noise is correlated with the current flow in the plasma. The emission occurs at the current front in a time on the order of 10 ns. We might note that the plasma impedance was purely resistive and that the current curve reproduced the shape of the voltage curve.

Figure 78 shows the emission spectrum of the turbulent plasma in the VIKhR'-2 and how it changes with the initial

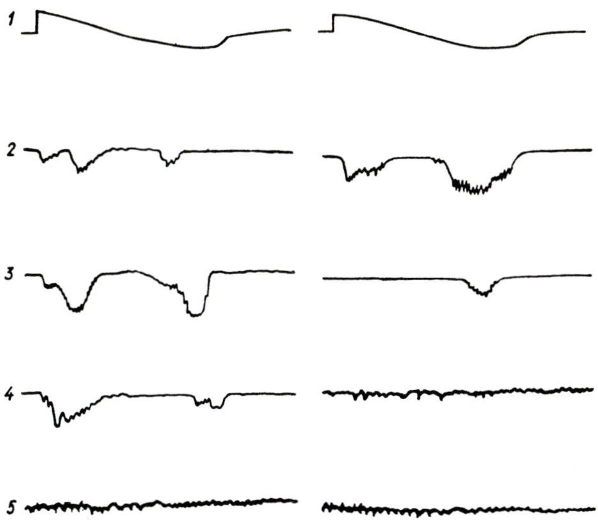

FIGURE 77 Traces of the superthermal microwave emission from the plasma in the VIKhR'-2 device[92,206] for two initial densities. *a*: $2\cdot10^{13}$ cm^{-3}. *b*: $2\cdot10^{12}$ cm^{-3}. *1*) Torus loop voltage, E = 250 V/cm; *2*) $\lambda \approx 30$; *3*) $\lambda \approx 15$; *4*) $\lambda \approx 8$; *5*) $\lambda \approx 4$ mm. The sweep time is 1 μs.

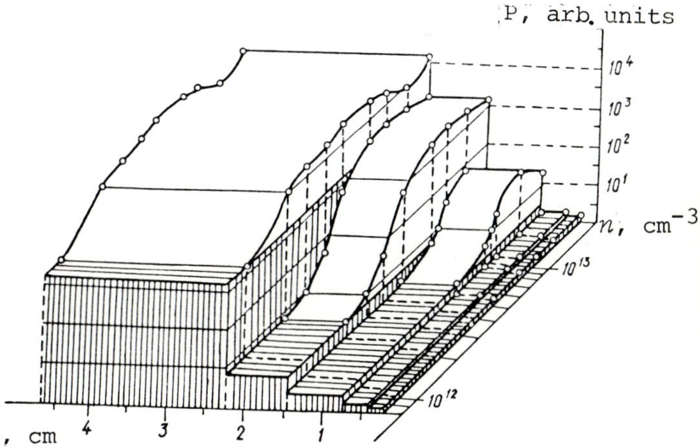

FIGURE 78 Histogram of the changes in the spectral intensity of the superthermal microwave emission from the turbulent plasma in the VIKhR'-2 device[206] 80 ns after the beginning of the heating as a function of the density ($n_{max} \simeq 2\cdot10^{13}$ cm^{-3}, E = 250 V/cm, H_{max} = 15 kOe). The hatching shows the receiver noise.

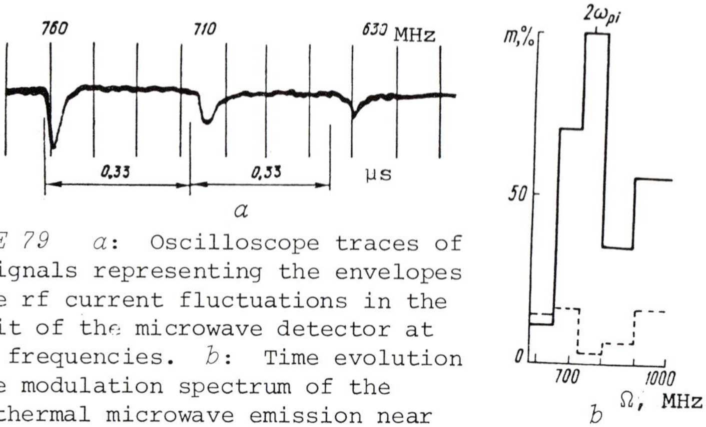

FIGURE 79 *a*: Oscilloscope traces of the signals representing the envelopes of the rf current fluctuations in the circuit of the microwave detector at three frequencies. *b*: Time evolution of the modulation spectrum of the superthermal microwave emission near $\omega \approx \omega_{pe}$ ($\lambda_{pe} \sim 19$ mm) in the VIKhR'-2 device.[142] Solid histogram) 20 ns; dashed histogram) 60 ns.

plasma density. The spectrum is plotted as a histogram for the time 80 ns after the current begins to flow; it corresponds to the spectrum of a fully developed current-driven turbulence, with a broad set of plasma frequencies, under conditions corresponding to a pronounced density inhomogeneity and a very bumpy magnetic field, $\frac{H_{max}}{H_{min}} = 20$. Here the emission power is averaged over the passband of each detector head, whose width is estimated to be roughly 50%. (Special measurements showed that the plasma density did not change by more than 10% during the short current pulse, 0.3 μs long). The emission intensity falls off sharply toward the short-wave boundary of the spectrum. The same effect is observed as the initial plasma density is reduced.

The superthermal microwave emission in the VIKhR'-2 contained a relatively narrow-band modulation spectrum near the ion plasma frequencies, $\sim 2\omega_{pi}$ (Ref. 142).

The traces in Fig. 79 are video signals of the envelopes of the rf current fluctuations in the circuit of the microwave detector at three frequencies: 760 ± 20, 710 ± 20, and 630 ± 30 MHz. These signals were recorded simultaneously by a three-channel rf analyzer during the flow of the plasma current (in a pulse 0.3 μs long). Also shown here are spectra of the modulation of the plasma-frequency microwave emission ($\lambda_{pe} \approx 1.9 \pm 0.1$ cm) for two times, 20 and 60 ns. We see that the fluctuations near the second harmonic of the ion plasma frequency, $\Omega \approx \frac{2\omega_{pi}}{2\pi} \approx 750$ MHz, occur with the minimum delay, at the front of the turbulent-heating current (the solid histogram). The modulation index near these frequencies is nearly 100%. As time elapses, the integrated microwave emission power increases by a factor of more than five (traces 1 and 3 in Fig. 77), and the broad fluctuation spectrum (dashed histogram) stretches down to extremely low frequencies, $0.1\omega_{pi} \sim 100$ MHz. The low-frequency wave spectrum in the plasma was also measured by detecting the Raman scattering of signals at $\lambda_0 \simeq 2$, 4, and 8 mm with magnetic and electric probes.

A deep modulation of the superthermal microwave emission from a plasma near ω_{pe} by a relatively narrow-band frequency spectrum, $\Omega \approx 2\omega_{pi}$, , was observed in the URAGAN-1 stellarator.[262] Measurements carried out in various gases showed that at a fixed plasma density the maximum in the modulation spectrum of the microwave emission at ω_{pe} depends on the ion mass: $\Omega \sim \sqrt{\frac{m_e}{M_i}}$. Furthermore, the modulation spectrum depends on the plasma density in such a manner that the following condition holds:

$$\frac{\Omega_1}{\omega^{(1)}} \simeq \frac{\Omega_2}{\omega^{(2)}} = \text{const} \approx 2\sqrt{\frac{m_e}{M_i}}.$$

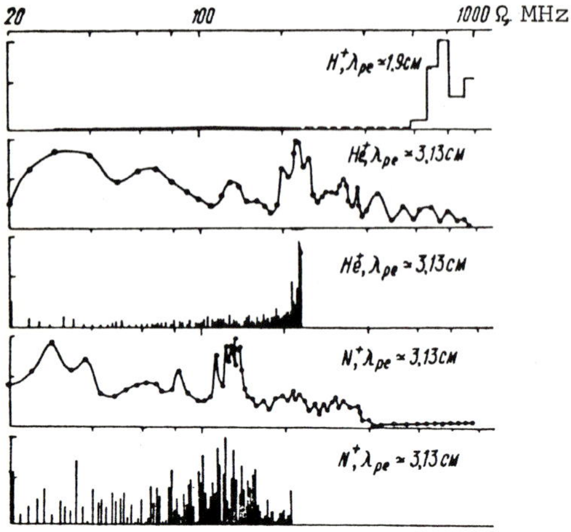

FIGURE 80 Changes in the modulation spectrum of the superthermal microwave emission at $\omega \simeq \omega_{pe}$ as a function of the ion mass in the URAGAN stellarator.[262] Shown at the top for comparison is a spectrum obtained in the VIKhR'-2 device.[142]

Figure 80 shows how the modulation spectrum of the superthermal microwave emission near the electron plasma frequency varies with the ion mass. These results were obtained through a simultaneous analysis of the spectrum of current fluctuations in the circuit of a broad-band square-law microwave detector $\left(\frac{\Delta\omega}{\omega} \approx 0.1\right)$ by four narrow-band superheterodyne receivers and through a Fourier spectral decomposition of the detector current. Shown for comparison in Fig. 80 is the modulation spectrum obtained in the VIKhR'-2 device.[142]

It may thus be said that there are common features in the microwave-emission mechanisms involving the nonlinear conversion of plasma waves in a turbulent plasma in the cases of strong electric fields ($E \gg E_{Dr}$; the VIKhR'-2 device[142]) and relatively weak fields ($E \tilde{>} E_{Dr}$; the URAGAN-1 device[262,263]). In both cases there is a strong interaction of a high-frequency (Langmuir) noise and a low-frequency ion-plasma noise. This interaction gives rise to the emission of electromagnetic waves near the plasma frequency. In a first approximation we can write the conservation laws

$$\omega_{pe}^{t}=\omega_{pe}^{l}+\Omega, \quad \mathbf{k}^{t}=\mathbf{k}^{l}+\mathbf{k}^{\Omega}$$

for these processes. These relations hold only in the case $\mathbf{k}^{\Omega}\ll\mathbf{k}_{D}$ at $\Omega\simeq\omega_{pi}$, $2\omega_{pi}$. These conditions may actually be satisfied in a turbulent plasma, since long-wave nonelectrostatic waves of the whistler type near harmonics of the ion plasma frequency have been detected by magnetic probes in the plasmas in turbulent-heating experiments in straight and toroidal discharges.[113,142] Furthermore, the fact that this noise is excited suggests that a magnetic field plays some role in the mechanism for the superthermal microwave emission.

Experiments carried out over a broad range of electric fields, $E \approx 0.1$-400 V/cm, in closed magnetic confinement systems show how the plasma magnetization factor $\frac{\omega_{ce}}{\omega_{pe}}$ affects the spectrum of the superthermal microwave emission.

A dynamic study of the microwave spectra over the frequency band 8-40 GHz was carried out in the $l = 2$ TWIST stellarator[224,226] with a ten-channel system of superheterodyne receivers. These experiments were carried out with very strong electric fields ($E \simeq 400$ V/cm at $\frac{\omega_{ce}}{\omega_{pe}} \cong 0.13$-1, $n \cong 5\cdot10^{11}$-$1.3\cdot10^{13}$ cm^{-3}, and $H = 1.6$-2.5 kOe). Figure 81 shows oscilloscope traces of various signals which characterize the highly turbulent discharge in the TWIST, for various values of the ratio $\frac{\omega_{ce}}{\omega_{pe}}$. Intense microwave signals appear after a delay of 150-200 ns with respect to the time at which the electric field is applied to the plasma loop, as in the VIKhR'-2 (Fig. 77). This delay is not a consequence of the response time of the detection system (the time constant of the video detectors does not exceed 10 ns; the passband of each channel is $\frac{\Delta\omega}{\omega} \simeq 1\%$).

Figure 82 shows the time evolution of the microwave signals during discharges in argon for various initial plasma densities. In a dense, unmagnetized plasma, with $\frac{\omega_{ce}}{\omega_{pe}} << 1$ (spectrum *a*), the emission peak initially (at 150 ns) is near the frequency $2\omega_{ce}$. Later on (at 170-230 ns), an intense peak arises near the plasma frequency ω_{pe}. At $\frac{\omega_{ce}}{\omega_{pe}} \simeq 1$ (spectrum *b*) the emission intensity is distributed nonuniformly, approximately in accordance with $P(\omega) \sim \omega^{-3}$, over the frequency range studied. In both cases the power level of the microwave emission is 10^2-10^3 times the thermal noise power with $T_e \cong 10^4$ eV.

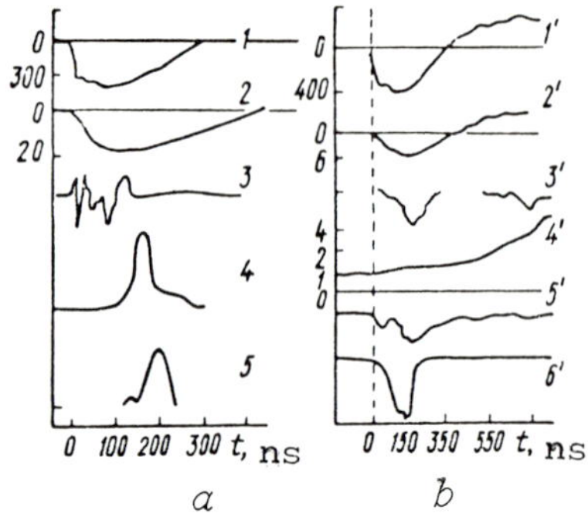

FIGURE 81 Oscilloscope traces from the TWIST stellator.[224,226] *a*: Argon discharge. *b*: Hydrogen discharges. *1*) Electric field; *2*) discharge current; *3*) signal from electrostatic probe; *4*) microwave emission at 9.6 GHz; *5*) microwave emission at the plasma frequency, 33.4 GHz (the initial parameters are $n^0_{max} = 1.3 \cdot 10^{13}$ cm^{-3}, $T_e \simeq 2$ eV, H = 1.6 kOe). *1'*) Electric field; *2'*) discharge current; *3'*) microwave emission; *4'*) plasma density n, $\times 10^{12}$ cm^{-3}; *5'*) scattering signal, $\lambda_0 = 2$ mm ($\Delta\omega \simeq 490$ MHz); *6'*) x-ray emission (H = 3 kOe).

Over the range of the magnetization parameter which

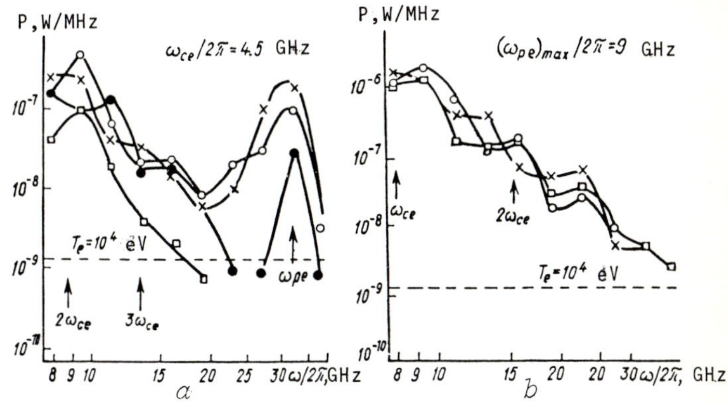

FIGURE 82 Time evolution of the spectrum of the superthermal microwave emission in the TWIST stellarator[226] for argon discharges at various values of the ratio ω_{ce}/ω_{pe}. *a:* $n = 1.3 \cdot 10^{13}$ cm^{-3}, $\omega_{pe}/2\pi$ = 33.6 GHz, $\omega_{ce}/2\pi$ = 4.5 GHz. *b:* $n = 1 \cdot 10^{12}$ cm^{-3}, $\omega_{pe}/2\pi$ = 9 GHz, $\omega_{ce}/2\pi$ = 8 GHz. Dashed lines) Level of the thermal microwave emission, $T_e \simeq 10^4$ eV; □) 150; o) 170; x) 200; ●) 230 ns.

was studied in the TWIST device, $\omega_{ce}/\omega_{pe} \approx 0.1$-1, no significant effect of the magnetic field on the nature of the emission was observed. Only by increasing the electric field was it possible to significantly increase the emission intensity and flatten the emission spectrum.

The superthermal microwave spectra were also studied over the frequency range 19-50 GHz in the TOR-2M and TOR-4M tokamaks,[23,24,26] in the corresponding range of the magnetization parameter, ω_{ce}/ω_{pe} = 0.3-0.8, during turbulent plasma heating by a short pulse (with a length on the order of 0.5 μs) of a loop electric field E = 40 V/cm. In the TOR-2M device,[23,25] with $\omega_{ce}/\omega_{pe} \cong 0.3$, the microwave spectrum was studied with interference filters, whose output signals were measured with a cooled InSb detector. Intense superthermal microwave emission with a spectral power density of about $1.5 \cdot 10^{-12}$ W/Hz was observed only near the plasma frequency.

A similar emission spectrum was obtained in experiments in the TOR-4M, in a single discharge pulse, through the use of a holographic Fourier spectrum analyzer with a resolution $\Delta\omega = 3.5$ GHz (Refs. 24 and 27). Figure 83 shows the spectrum and the output signals from the analyzer. The interference pattern formed in the region where the two wave trains were superimposed in the holographic analyzer was detected with eight wide-band detectors. The output signals from these detectors were fed to an oscilloscope through delay lines at intervals of 1 μs (Fig. 83a). The nature of the spectrum (Fig. 83b) at $\omega_{ce}/\omega_{pe} = 0.5$-$0.8$ is qualitative evidence for the excitation of intense plasma waves and ion acoustic waves in strong electric fields, $E \gg E_{Dr}$.

The effect of the magnetization factor on the microwave emission spectrum could be seen particularly clearly in approximately steady toroidal discharges, where turbulent plasma heating by a current was (generally) carried out near the threshold, $E/E_{Dr} \gtrsim 1$.

Experiments carried out in the URAGAN-2 stellarator with current heating over the broad range $\omega_{ce}/\omega_{pe} = 0.5$-$3$ and with $E = 0.01$-0.2 V/cm (Refs. 263, 264, and 52-54), showed that the transition of a turbulent plasma with an anomalous ion acoustic resistance to a regime with sporadic superthermal microwave emission near the frequency $\omega \approx \omega_{pe}$ depends exclusively on the strength of the magnetic field. The plasma parameters were monitored by a large array of diagnostic instruments. In particular, the average density and the maximum density were measured by phase methods at the wavelengths $\lambda_0 = 2$ and 8 mm, and they were also measured by a laser method. The electron temperature was determined from the thermal microwave emission, from the

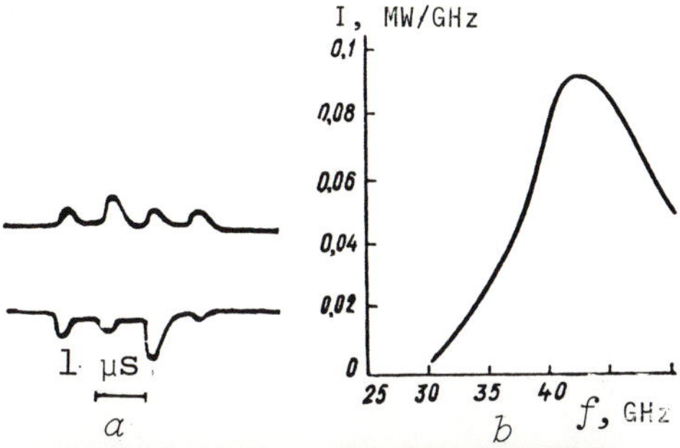

FIGURE 83 *a:* Output signal from the holographic Fourier spectrum analyzer. *b:* spectrum of the superthermal microwave emission from the plasma in the TOR-4M device[24,27] ($n = 2.2 \cdot 10^{13}$ cm^{-3}, $\omega_{ce}/\omega_{pe} = 0.5$, $E \simeq 40$ V/cm).

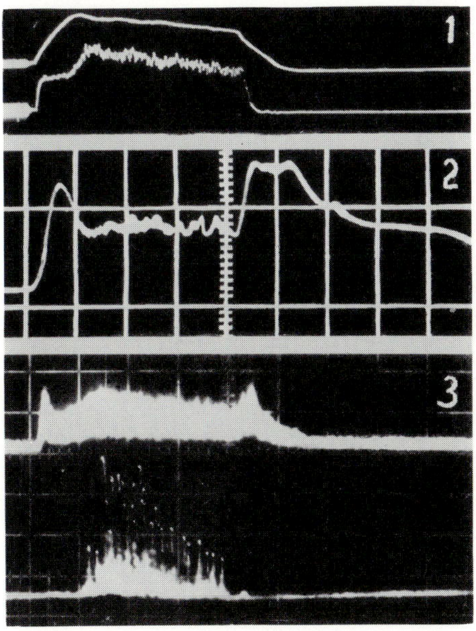

FIGURE 84 Oscilloscope traces illustrating the helium discharge in the URAGAN-1 stellarator[78,52,53] ($H = 4.8$ kOe). *1*) Current of 3.2 kA, electric field of 0.03 V/cm; *2*) plasma density (the scale is $4 \cdot 10^{12}$ cm^{-3}/div); *3*) thermal microwave emission at $\lambda = 1.6$ cm and superthermal microwave emission at $\lambda = 3.1$ cm (the sweep time is 1 ms/div).

Thomson scattering of a laser beam, and by spectroscopic and diamagnetic methods.[78,80]

In accordance with the plasma parameters ($n = 2 \cdot 10^{12}$–$2 \cdot 10^{13}$ cm^{-3}, $H = 4.8$–19.2 kOe), the microwave spectra were studied over the wavelength range $\lambda_0 = 30$–4 mm with a six-channel system of superheterodyne receivers with a threshold sensitivity over the range 0.1-1 eV and with microwave filters having a variable passband ($\Delta\omega = 1$ GHz, 20 MHz). The video signals were displayed by linear and logarithmic

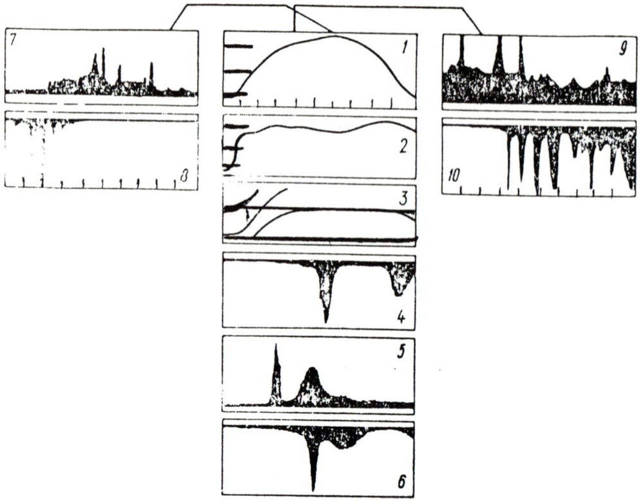

FIGURE 85 Helium discharge in the URAGAN-2 stellarator[53] (H = 16.8 kOe). *1*) Current in the plasma, 500 A/div; *2*) electric field, 0.005 V/cm; *3*) average plasma density, $2 \cdot 10^{12}$ cm^{-3}/div; *4*) hard x-ray emission in the energy range 0.5-1 MeV; *5,6*) emission intensities at the plasma frequencies, 15.7 and 25.4 GHz (the sweep time is 1 ms/div); *7,8*) hard x-ray emission in the energy range 0.5-1 MeV and emission intensity at 19.8 GHz (sweep time of 0.5 ms/div); *9,10*) emission intensities at 15.2 and 19.1 GHz (50 μs/div).

systems over a dynamic range of 40 dB with a time constant of 0.1 μs. The microwave measurement system was calibrated with a standard wide-band noise source using gas-discharge tubes; the geometry of the stellarator discharge chamber and of the antenna-feeder system was taken into account.[53] Figures 84 and 85 are oscilloscope traces of the current discharge in helium for two values of the magnetic field: 4.8 and 16.8 kOe. The plasma emits a steady-state thermal microwave noise for a time interval which correlates with the current, and it also emits intense sporadic microwave signals.[54]

The spectral power density is extremely nonuniform over the wavelength range studied. The nature of the spectrum changes as the magnetic field is strengthened. Figure 86 shows spectra for three values of the magnetic field (4.8, 7.2, and 16.8 kOe), for a constant, controlled plasma density in the stellarator, $n^{max} \simeq 4\cdot10^{12}$ cm^{-3}. It was found that in strong magnetic fields the emission near the plasma frequency ω_{pe}^{max}, becomes nonthermal and time-varying; its power level exceeds the thermal power level by three or four orders of magnitude.

Three basic discharge regimes, differing in the nature of the microwave emission near the plasma frequency, were identified for the central part of the plasma column in the URAGAN-2 stellarator:

1) $E/E_{Dr} \gtrless 1$, $\omega_{ce}/\omega_{pe} \lesssim 1$ (the high-temperature regime),
2) $E/E_{Dr} \ll 1$, $\omega_{ce}/\omega_{pe} > 1$ (the acceleration regime),
3) $E/E_{Dr} > 1$, $\omega_{ce}/\omega_{pe} > 1$ (the turbulent regime).

In weak magnetic fields, corresponding to regime 1, the brightnesss temperature of the steady-state microwave noise behaves in the same way as the electron temperature T_e over a broad range of electric fields, in both turbulent discharges and discharges with a classical conductivity. The electron temperature T_e measured by a radiometer method agreed with the temperature determined from measurements of the scattering of the beam from a ruby laser.[78,80] The spectral peak of the thermal microwave

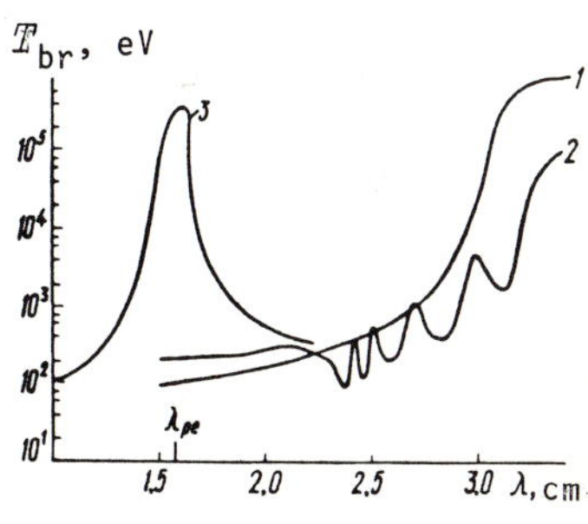

FIGURE 86 Microwave emission spectra in the URAGAN stellarator[53,264] as a function of the magnetic field at a constant density (helium). $n = 4.2\cdot10^{12}$ cm^{-3}, $E = 1.1\cdot10^{-2}$ V/cm, *1*) $H = 4.8$; *2*) $H = 7.2$; *3*) $H = 16.8$ kOe.

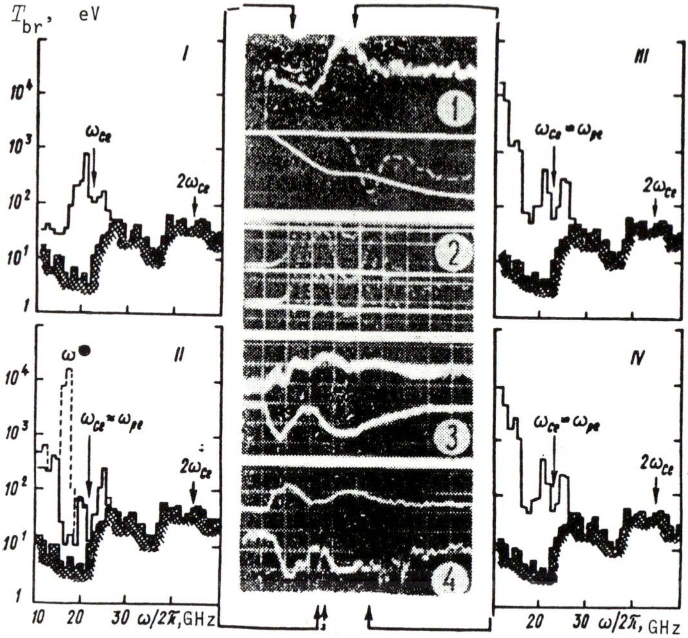

FIGURE 87 Time evolution of the microwave spectra of the turbulent hydrogen plasma in strong electric fields in the URAGAN-2 stellarator[54] ($\omega_{ce}/\omega_{pe} \approx 1$, $H = 7.85$ kOe). Oscilloscope traces: *1*) Electric field E, 0.05 (V/cm)/div, discharge current, 3.8 kA/div, the impurity line CIII 4647 Å (dashed curve); *2*) average density, $2 \cdot 10^{12}$ cm^{-3} between lines; *3*,*4*) microwave emission at the frequencies 26, 21 and 23, 18 GHz, respectively, (the vertical scales are logarithmic, spanning two or three orders of magnitude; the horizontal scale is 100 μs/div). The spectra were obtained at the following times: *I*) 100 μs, *II*) 220 μs (solid histogram) and 240 μs (dashed histogram) ($\omega^{\bullet}$ is a sporadic emission burst, $\omega^{\bullet}_{pe} < \omega^{max}_{pe}$); *III*) 350 μs; *IV*) 450 μs after the beginning of the discharge. The cross hatching shows the threshold sensitivity of the receiver system.

emission occurred near the plasma frequency, $\omega_{max} \approx \omega^{max}_{pe}$.

Intense superthermal microwave emission occurred in regime 1 only at frequencies $\omega^{\bullet} < \omega_{ce}$, at characteristic times corresponding to the beginning of the drop in the plasma density, similar to the density drop which had been observed previously in the model B1 stellarator.[212] We might note that the experimental results which we will be discussing here correspond to only the "pure" stages of the discharge in the URAGAN-2 stellarator, over the time interval 250-400 μs, during which we can completely rule out the effects of an impurity of heavy ions and a secondary gas from the wall of the discharge chamber.

Figure 87 shows the time evolution of the microwave spectra for the initial stage of a highly turbulent hydrogen discharge in a weak magnetic field, $\omega_{ce}/\omega_{pe} \approx 1$, and in a strong electric field, $E/E_{Dr} \approx 6$. In this case the current velocity stabilizes at $u \approx 8v_S$, and the observed resistance anomaly is not a consequence of an increase in the effective charge caused by a heavy-ion impurity. At the gas-ionization front the spectrum has a narrow peak near the cyclotron frequency, ω_{ce}, with a characteristic "red shift" (spectrum I). At the density maximum (spectrum II, solid histogram) there are two peaks near frequencies $\omega_{ce} \approx \omega_{pe}$: a "violet" peak with a brightness temperature $T_{br}^{viol} \approx 300$ eV, which is higher than the transverse electron temperature, $T_e^{\perp} \approx 50$ eV measured by the laser; and a "red" peak, with a temperature near that found by the laser measurements, $T_{br}^{red} = T_e^{\perp} \simeq 50$ eV, (see Fig. 68). These facts may indicate a pronounced anisotropy of the temperature ($T_e^{\parallel} >> T_e^{\perp}$) in the strong electric fields. The dashed histogram in spectrum II in Fig. 87 is the spectrum of an intense, narrow-band, sporadic burst at $\omega^{\bullet} < \omega_{ce} \approx \omega_{pe}^{max}$, which lasts 15 μs.

The burst is seen to be preceded by a drop in the plasma density and by the appearance of impurity lines, e.g., CIII 4647 Å.

Probe measurements and the nature of the observed microwave spectra showed that the frequencies of the sporadic bursts corresponded to plasma frequencies excited in the less dense plasma at the periphery of the column,[264,132] $\omega_{pe} < \omega_{ce}$.

During the next 150-200 μs after the microwave burst and the density drop (traces 2 and 4 in Fig. 87), the emission spectra show a pronounced intensification at low frequencies and a characteristic splitting of the line near the frequency $\omega_{ce} \simeq \omega_{pe}^{max}$ into red and violet components (spectra III and IV), with approximately equal brightness temperatures, $T_{br}^{red} \cong T_{br}^{viol} \approx 300\text{-}450$ eV.

The appearance of the sporadic microwave bursts under these conditions in the stellarator could not be explained by the classical, approximately steady runaway of electrons in a field $E > E_{Dr}$. The experiments revealed a sharp disruption of the electron acceleration in the energy range 20-500 keV with increasing electric field, beginning at the threshold $E/E_{Dr} \simeq 0.1$; this effect was apparently caused by an effective scattering of superthermal electrons in the turbulent plasma[80,79] (see Fig. 68). Special measurements with composite microwave and x-ray probes in the plasma showed that the burst of the hard x-ray component (corresponding to energies up to 10 keV) from the target arose in a bounded region in the plasma at the same time as the microwave signal and was of the nature of a sporadic pulse.[264]

In strong magnetic fields ($\omega_{ce}/\omega_{pe} > 1$) and weak electric fields ($E/E_{Dr} << 1$) (regime 2), the discharges in the URAGAN stellarator had a large number of fast electrons. The energy measured for the hard x rays was about 0.5 MeV

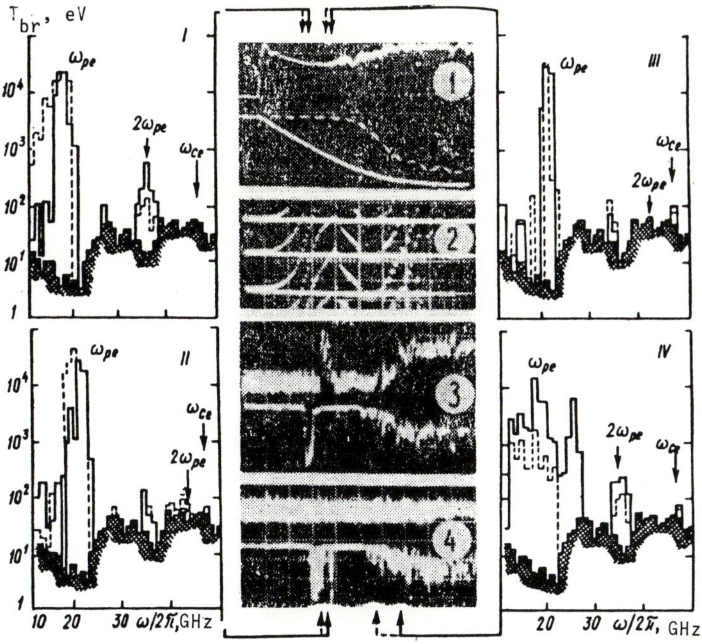

FIGURE 88 Time evolution of the microwave emission spectra of a turbulent hydrogen plasma in the URAGAN-2 stellarator[54] in a strong magnetic field, $\omega_{ce}/\omega_{pe}^{max} \approx 2.2$, $H = 16.8$ kOe. Oscilloscope traces: *1*) Electric field E, 0.05 (V/cm)/div, discharge current, 3.8 kA/div, the impurity line CIII 4647 Å (dashed curve); *2*) average density, $2 \cdot 10^{12}$ cm^{-3} between lines; *3,4*) microwave emission at the frequencies 23, 17, 28, and 22 GHz, respectively (the vertical scale is logarithmic, spanning three or four orders of magnitude; the horizontal scale is 100 μs/div). The spectra were recorded at the following times: *I*) 180 μs (dashed histogram), 200 μs (solid histogram); *II*) 320 μs (dashed histogram), 350 μs (solid histogram); *III*) 370 μs (dashed histogram), 390 μs (solid histogram); *IV*) 450 μs (dashed histogram), 550 μs (solid histogram); cross hatching) threshold sensitivity of the receiver system.

and corresponded to calculated energy of electrons accelerated in an approximately steady electric field* (see traces

*O. S. Pavlichenko, F. F. Tereshchenko, and V. M. Tonkopryad, Fiz. Plazmy 3, No. 2, 321-328 (1977) [Sov. J. Plasma Phys. 3, 180 (1977)]

4 and 10 in Fig. 85). In this case the intensity and spectrum of the microwave emission corresponded to the acceleration regimes observed in the Model B1 stellarator,[192,230] the S-1 stellarator,[37] the TM-6 tokamak,[296] the ALCATOR tokamak,[204] and the TFR tokamak.[292,215]

The transition to the highly turbulent regime (regime 3), with $E/E_{Dr} > 1$ and $\omega_{ce}/\omega_{pe} > 1$, was accompanied by a disruption of the quasisteady electron acceleration and by a corresponding decrease in the microwave emission intensity, by two orders of magnitude. A superthermal microwave emission near the maximum plasma frequency was observed in this case in the form of sporadic short bursts.

Figure 88 shows the time evolution of the microwave spectra in the highly turbulent regime (regime 3) in the URAGAN-2 stellarator.[54] We see that the intense, sporadic, narrow-band bursts at the plasma frequency occur primarily near the density maximum, at 300-350 μs. The signals at $2\omega_{pe}$, ω_{ce}, and $2\omega_{ce}$ are greatly weakened. The intensity of the emission at the cyclotron harmonics is at the thermal noise level.

The characteristic cyclotron emission from the plasma in the turbulent regime is related to an anomalous spectral broadening,* which is much greater than either the thermal or magnetic broadening, for example, in a straight section of the URAGAN stellarator, with $\Delta H/H \approx 1.4\%$.

---

*Turbulent broadening of the cyclotron spectrum in a linear discharge has been observed by H. Iguchi, Y. Yto, T. Kawabe, A. Mase, and T. Tsukisima ["Observations of cyclotron radiations from turbulently heated plasma (THE MACH II)," Phys. Lett. 62A, No. 5, pp. 343-346 (1977)].

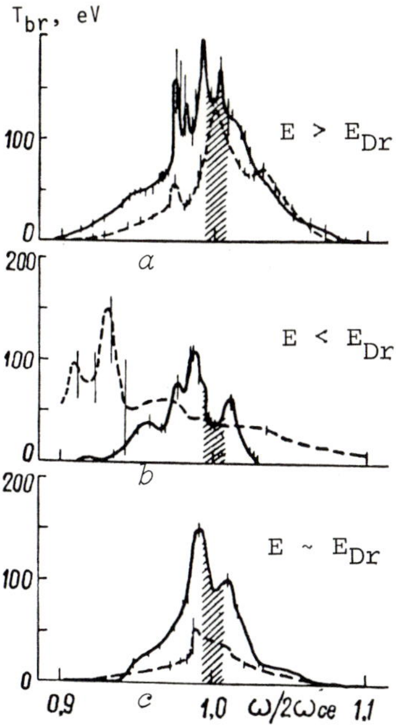

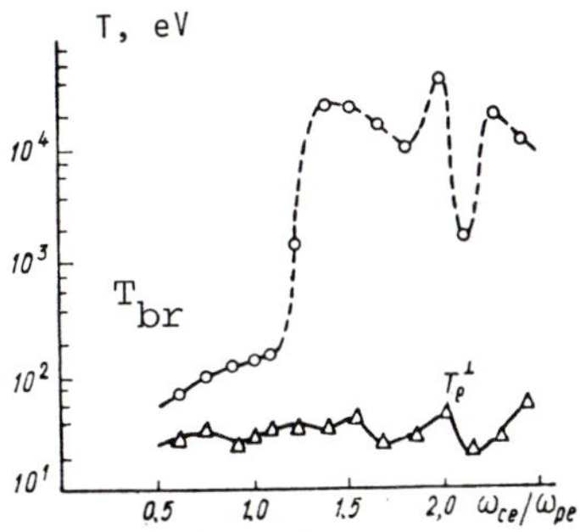

FIGURE 89 Evolution of the emission spectra near the second harmonic of the electron cyclotron frequency, $2\omega_{ce}$, with decreasing electric field in the URAGAN-2 stellarator at a constant magnetic field, $H \approx 13$ kOe, in hydrogen (dashed curves) and deuterium (solid curves). *a*) $E = 0.1$ and *b*) $E = 1.5 \cdot 10^{-2}$ for $n = 6 \cdot 10^{12}$ cm^{-3}; *c*) $E = (1\text{-}2) \cdot 10^{-2}$ V/cm for $n = 3 \cdot 10^{12}$ cm^{-3}. The hatching shows the broadening due to the inhomogeneity of the magnetic field in the straight region in the stellarator.

FIGURE 90 Threshold in the magnetic-field dependence of the brightness temperature of the plasma at the electron plasma frequency; magnetic-field dependence of the transverse electron temperature in a turbulent hydrogen discharge in the URAGAN-2 stellarator.[54] $E = 0.1$ V/cm, $n = 6 \cdot 10^{12}$ cm^{-3}.

Figure 89 shows the evolution of the spectrum at the cyclotron harmonic $2\omega_{ce}$ as the electric field is raised in hydrogen and deuterium discharges. In regime 3, in which intense microwave emission is observed at ω_{pe}, and in which there is no steady electron acceleration, the cyclotron emission has a Raman spectrum (Fig. 89a), with a typical splitting of the peak into several satellites. The spacing of the satellites depends on the mass of the gas $[\Delta\omega \sim (M_i)^{-1/2}]$.

In regime 2, in which intense hard x radiation is detected, the cyclotron-emission spectrum (Fig. 89b) exhibits a relativistic red shift, which corresponds to an x-ray energy on the order of 0.05 MeV. The time-varying nature of the synchrotron spectrum under conditions such that the velocity of the runaway electrons is much higher than the thermal velocity, $v \sim 0.2c \gg \frac{\omega_{ce}}{\omega_{pe}} v_{Te}$, suggests an instability due to the anomalous Doppler effect.[241,140] In a stronger electric field, in the absence of free runaway electrons, the noise intensity near the center of the resonant-absorption line (Fig. 89c) is determined by the electron temperature.[205,190,293]

Analysis of the microwave emission spectrum of a magnetized plasma over a broad frequency range thus suggests a primarily turbulent mechanism for the sporadic superthermal emission at ω_{pe} and suggests that this emission is not a consequence of free runaway electrons in strong electric fields,* $E > E_{Dr}$. Several other pieces of evidence obtained in experiments in the URAGAN-2 stellarator[262] contribute to an understanding of the time-varying, narrow-band plasma spectra.

---

*A corresponding conclusion has been reached for tokamaks by Y. Mok and C. S. Liu ["Theory of nonthermal radiation at plasma frequency in tokamaks," Bull. Am. Phys. Soc. 22, No. 9, 1201 (1977)].

Figure 90 shows the dependence of the brightness temperature of the emission at the plasma frequency on the magnetic field for the highly turbulent regime (regime 3). With increasing magnetic field, sporadic superthermal signals are excited at ω_{pe}^{max} (the dashed curve) near the threshold, $\omega_{ce}/\omega_{pe} \approx 1$. It has been shown[262,132] that the primary cause of this effect is a change in the dispersive properties of the magnetized plasma for the wave branch excited by the current in the plasma near the lower hybrid resonance. The intensity of this noise increases sharply near the threshold, causing an effective heating of the ions (Fig. 49). There is also an increase in the plasma conductivity; the drift velocity stabilizes at $u \approx 0.5v_{Te}$ (Fig. 69). We might note that the results in Fig. 90 were obtained at a constant plasma density and at a constant value $E/E_{Dr} \sim 6$, in the absence of free runaway electrons.

From the ratio of the power levels in the superthermal plasma harmonics, $P_{2\omega_{pe}}/P_{\omega_{pe}}$ (Fig. 88), and from Eqs. (VI.2) and (VI.3), we can estimate the energy redistribution in the plasma-wave (W_l) and ion acoustic (W_i) spectra in the turbulent plasma. Figure 91 shows the ratio W_l/W_i and several other plasma parameters, plotted as functions of the electric field for the acceleration and turbulent regimes in a single discharge in the URAGAN stellarator. In the acceleration regime the decrease in W_l/W_i with increasing electric field correlates with the decrease in the x-ray emission intensity (curves 3 and 4). The intensification of the ion-plasma-wave noise (2) with increasing field causes a stabilization of the ratio at $W_l/W_i \sim 0.1$ (5) and directly determines the conditions for the nonlinear conversion of plasma waves in the plasma.

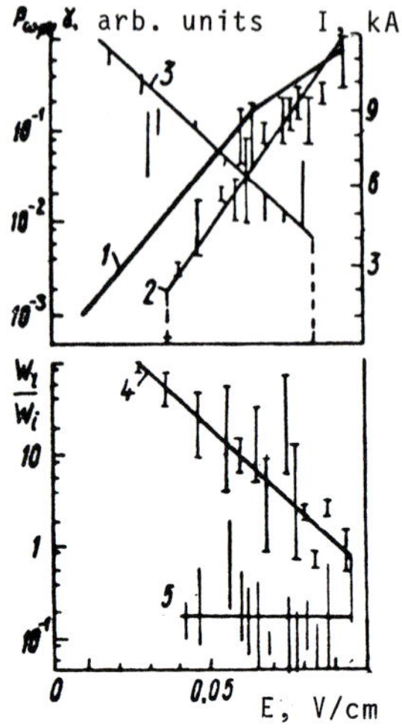

FIGURE 91 Ratio of the energy densities of the electron plasma waves and the ion plasma waves as a function of the electric field in the URAGAN-2 in the acceleration stage of the discharge (t = 200 μs) and in the turbulent stage (t = 300-400 μs). *1*) Current in the turbulent stage, H = 16.8 kOe, $n = 6 \cdot 10^{12}$ cm^{-3}, $\omega_{ce}/\omega_{pe} \approx 2.2$; *2,5*) intensity of the ion plasma noise, $P_{\omega pi}$, and the ratio W_l/W_i in the highly turbulent stage, in which there is no approximately steady acceleration of electrons; *3,4*) intensity of the hard x-ray emission, γ (up to 30 keV), and the ratio W_l/W_i in the acceleration stage. The dashed lines show the thresholds for the cutoff of the acceleration (*3*) and the threshold for the excitation of ion turbulence (*2*).

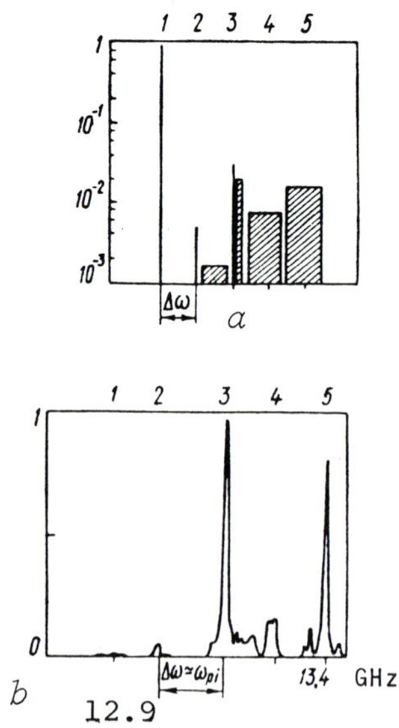

FIGURE 92 Spectral fine structure of a sporadic microwave burst near the plasma frequency in the URAGAN-2 (Ref. 264). *a*: Analysis of the modulation spectrum of the microwave emission, $\omega_{pe}/2\pi = 12 \pm 0.5$ GHz. *1*) $\omega_{pe} + 2\omega_{pi}$; *2*) $\omega_{pe} + 3\omega_{pi}$; *3*) $\omega_{pe} + 4\omega_{pi}$; *4*) $\omega_{pe} + 5\omega_{pi}$; *5*) $\omega_{pe} + 6\omega_{pi}$ (H_0 = 4.8 kOe). *b*: Output spectrum from the fast microwave analyzer (analysis time, 5 μs), $\omega_{pe}/2\pi = 13.1 \pm 0.35$ GHz. *1*) ω_{pe}; *2*) $\omega_{pe} + \omega_{pi}$; *3*) $\omega_{pe} + 2\omega_{pi}$; *4*) $\omega_{pe} + 3\omega_{pi}$; *5*) $\omega_{pe} + 4\omega_{pi}$ (H_0 = 7.2 kOe).

The spectral fine structure of the sporadic microwave bursts were studied in experiments in the URAGAN-2. It was found that in the turbulent current discharge the brief and intense microwave burst has a line spectrum. The emission occurs near the maximum electron plasma frequency, and the lines have a characteristic doublet structure with a separation equal to twice the ion plasma frequency, $2\Delta\omega \approx 2\omega_{pi}$. The quality factor of the emission lines, which contain up to 80% of the total energy of the microwave burst, is on the order of 10^3.

Figure 92 shows the spectral fine structure of a burst at the plasma frequency, ω_{pe}, lasting 15-20 μs for a helium plasma. Spectrum *a* is the product of a computer analysis of a modulation spectrum which is the statistical average over many pulses[263]; spectrum *b* was obtained in a single burst with the help of a fast microwave analyzer.[264]

A composite probe in the plasma in the URAGAN stellarator revealed an extremely small zone in which intense sporadic noise was excited near the frequencies ω_{pe}, ω_{pi} and $2\omega_{pi}$. In the absence of an approximately steady acceleration of electrons in this region, ejections of superthermal electrons and a large and abrupt decrease in the plasma density and in the magnetic field were observed simultaneously (Fig. 93). The spatial position in which these perturbation signals were localized depended on the relationship between the longitudinal and helical magnetic fields in the stellarator. The dimensions of the zone were much smaller than the cross-sectional area of the plasma column.

Analysis of the URAGAN results leads to the conclusion that an explosive release of wave energy — the energy of waves near the lower hybrid resonance, which had built up in cavitons — is the primary cause of the particle acceleration and of the emission bursts near the plasma frequency with a Raman spectrum, $\omega_{pe} + 2\omega_{pi}$ and $\omega_{pe} + 4\omega_{pi}$, during the evolution stage of a turbulent current discharge in a strong magnetic field. It was found that the confining magnetic

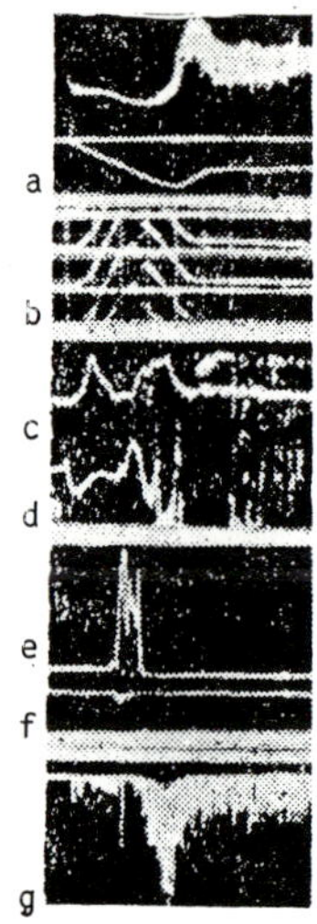

FIGURE 93 Oscilloscope traces for a hydrogen discharge in the URAGAN-2 stellarator[264] demonstrating the explosive nature of the emission and of the magnetic perturbations, detected simultaneously in the plasma by a composite probe. *a*) Loop voltage, 20 V/div, discharge current, 2 kA/div; *b*) plasma density ($2 \cdot 10^{12}$ cm^{-3} between lines); *c*) interferogram of the 2-mm signals; *d*,*e*,*f*,*g*) probe signals $\Delta H \approx 10^{-2} H$, microwave emission, $\omega \approx \omega_{pe}$, x-ray emission at energies below 10 keV, and noise at $\omega \sim 2\omega_{pi}$ (the horizontal scale is 250 μs/div; H = 7.2 kOe).

field serves as a reservoir of rf electromagnetic energy in the turbulent plasma, and it was shown that the ion plasma turbulence plays a governing role in the mechanism for the sporadic superthermal microwave emission.

An extremely important result, supporting these conclusions, is the estimated relative level of Langmuir and ion acoustic noise, W_l/W_i, which was found in the URAGAN stellarator and also in experiments on turbulent plasma heating in the TOR-4M device.[5,93] In the latter device, for example, these estimates were based on independent measurements of the energy density of the ion acoustic waves, W_i, from the Stark broadening of the hydrogen spectral line $H\gamma$ and the power level of the plasma-frequency emission, $P_{\omega pe}$. An estimate was made from Eq. (VI.2) under the assumption of a uniform wave-vector distribution of the wave energy from $\mathbf{k}=0$ to $\mathbf{k}=\mathbf{k}_0$. In a plasma with $n = 4\cdot 10^{13}$ cm^{-3}, H = 12 kOe, E = 40 V/cm, and T_e = 200 eV, the energy density of the ion acoustic waves was found to be substantially higher than the plasma-wave energy.density: $W_i \approx 10W_l$. The degree of ion turbulence was $W_i/nT \approx 2\cdot 10^{-3}$, and the field amplitude of the fluctuations reached $\tilde{E}_S \approx$ 6 kV/cm.

The model of the coalescence of plasma waves and ion acoustic waves trapped in cavitons was used by Rutgers and De Kluiver[274] to calculate the emission spectrum near harmonics of the electron plasma frequency and for a comparison of the results with the spectrum of plasma harmonics in the TURHE turbulent straight discharge.[277] These experiments yielded the most comprehensive information about the superthermal explosive spectrum of plasma-wave harmonics $\omega \geq \omega_{pe}$ in the case of very strong electric fields ($E \approx$ 500-800 V/cm) in a dense plasma, with $\omega_{ce}/\omega_{pe} \approx 0.5$ ($n = 2\cdot 10^{13}$-$2\cdot 10^{14}$ cm^{-3}). The emission was detected with a cooled InSb detector, and the spectrum was analyzed with wire-grating bandpass filters. The most intense microwave burst occurred near the current maximum, 1.6-2.0 μs after the beginning of the discharge, as in the experiments in the GROM-3 (Fig. 9; see also Ref. 137). The drift velocity was approximately equal to the ion acoustic velocity: $u \sim v_S$. The microwave burst was accompanied by a burst of hard x radiation.[274,277]

Figure 94 shows oscilloscope traces of the integrated power radiated in the submillimeter range from a turbulent hydrogen discharge and the spectrum of a microwave emission burst, 0.3 μs long, for two values of the plasma density. In this spectrum we see higher harmonics of the electron plasma frequency, $(1\text{-}6)\omega_{pe}$.

The spectrum was calculated from the model of Ikezi et al.,[237] in which there is a coalescence of solitary plasma waves and ion acoustic waves trapped in a density cavity with dimensions on the order of ten times the Debye wavelength, $(10\text{-}20)\lambda_D$. The open circles in Fig. 94 demonstrate the quantitative agreement between the calculated spectrum of plasma-wave harmonics and the spectrum observed in the TURHE device. Agreement was reached here by adopting the value $\frac{\delta n}{n} = \left(\frac{\tilde{E}^2}{4nT}\right) = 0.425$ as a measure of the displacement of plasma density from a cavity. This value shows that for discharge temperatures $T_e \sim 6$ keV and $T_i \sim 2$ keV the electric field of the plasma-wave fluctuations can reach $\tilde{E}_l \approx 500$ kV/cm.

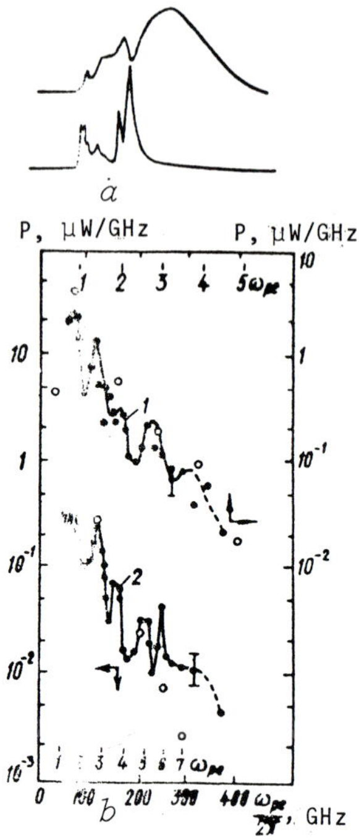

FIGURE 94 *a*: Traces of the discharge current and of the integrated microwave emission power (horizontal scale 1 μs/div). *b*: Plasma-harmonic spectrum in the TURHE device[274] at the time of a microwave burst (t = 1.6-2 μs). 1) ω_{pe} = 80 GHz, $n = 8\cdot10^{13}$ cm^{-3}; 2) ω_{pe} = 42 GHz, $n = 2.5\cdot10^{13}$ cm^{-3} (hydrogen, T_e = 6-10 keV, T_i = 2-3 keV, I = 4.5 kA, E = 400 V/cm, H = 8 kOe). ●) Experiment; o) theory.

Estimates based on the satellites of the forbidden helium lines $2^1P\text{–}4^1F$, λ = 4920 Å, put the electric fields of the fluctuations in the TURHE at $E_{ls} \sim 10$ kV/cm (Ref. 274). According to the theory of Sholin[182] and Griem,[220] the non-adiabatic Stark broadening of the central component of the hydrogen $H\alpha$ line can be used to determine the degree of Langmuir turbulence in the TURHE under the conditions specified above. These calculations yield $W_l/nT \simeq 0.1$-0.2 and, correspondingly, $\tilde{E}_l \approx 500$ kV/cm, in good agreement with the results calculated for the spectrum of plasma-wave harmonics. Such a high level of plasma waves in a turbulent straight discharge has been detected only in the TURHE device in very strong electric fields, $E \cong 800$ V/cm.

Optical Stark-effect measurements of the turbulent noise level in the NPR-2 device revealed a fluctuation field $\tilde{E}_l \approx 20$ kV/cm at a turbulent-heating electric field $E \simeq 250$ V/cm and with a degree of turbulence[94,93] $\frac{W_l}{nT} \sim 2\cdot 10^{-3}$.

In a straight discharge of the GROM type (Zelenin et al.[100]), with $n \simeq 5\cdot 10^{14}$ cm^{-3}, $T_e \simeq 10^3$ eV, and $E \simeq 500$ V/cm, the degree of turbulence was determined from measurements of satellites of the forbidden lines of helium, HeI 4470.03 Å ($2p\,^3P^0$–$4f\,^3F$): $W_l/nT \simeq 10^{-4}$.

Analysis of the experimental data from a variety of plasma devices over broad ranges of the plasma parameters shows that changes in the turbulent-heating conditions can cause a transition from the classical beam–plasma interaction mechanisms (in which the excitation of plasma waves and the superthermal microwave emission result from electron runaway in strong electric fields) to turbulent mechanisms, in which there may be a complex energy conversion in the ion acoustic spectrum[269,118,294] and particle trapping by the fields of the turbulent fluctuations (see the computer simulations by Drummond et al.[211] and by Biskamp and Chodura[194,196]).

The modulational instability[60] and caviton explosions are regarded as among the most likely mechanisms for the sporadic microwave emission at the plasma frequency in high-current discharges.[264,272] A model in which three-dimensional ordinary-wave solitons build up has been proposed by Korolev and Petviashvili[124] in an effort to explain, for example, type I solar radio bursts.

4 TURBULENT NOISE UNDER ANOMALOUS-RESISTANCE CONDITIONS

The frequency spectra of the superthermal turbulent fluctuations with $\omega \sim \omega_{pi}$ which were detected by probes in the plasmas of straight and toroidal discharges in experiments by Zavoĭskiĭ et al.[93] confirm the Rudakov—Stringer concept of the anomalous resistance. Furthermore, the observation of intense harmonics of the ion plasma frequency, $\omega \sim \omega_{pi}$ and $2\omega_{pi}$, with a magnetic probe[113] (and with electric and magnetic probes simultaneously) and for further observation of these harmonics in the modulation spectrum of the superthermal microwave emission[142] indicate that the turbulent fluctuations are nonelectrostatic, long-wave fluctuations.

Rudakov[113] has offered an explanation of the nonelectrostatic nature of the turbulent noise and of the mechanism for the excitation of harmonics of the ion plasma frequency in a nonisothermal plasma with $T_e >> T_i$ under anomalous-resistance conditions. His explanation is based on the possibility of a nonlinear conversion of short, electrostatic ion acoustic waves into electromagnetic whistlers. If the spectrum of the turbulent fluctuations excited in the plasma by the current has an energy maximum at ω_{pi}, then whistlers should be observed in a narrow spectral interval near the frequency $\Omega \sim 2\omega_{pi}$, and there should be a broader peak at $\Omega \tilde{<} \omega_{pi}$. It must be noted that at $\Omega/k > v_{Te}$ the whistlers, which propagate preferentially along the magnetic field, are damped only slightly.

As mentioned earlier, there are many nonintersecting natural wave branches in a magnetized plasma at frequencies near $\omega_{ci} < \omega < \omega_{ce}\cos\theta$. The dispersion properties of these waves have been analyzed in detail in the theoretical work by Ginzburg and Rukhadze.[63] For example, wave branches near the lower hybrid resonance,

$$\omega_L^2 = \omega_{ce}\omega_{ci}\left(1+\frac{\omega_{ce}^2}{\omega_{pe}^2}\right)^{-1},$$

have been studied thoroughly and are used for rf plasma heating.[64,70,151] In a strong magnetic field, $\omega_{ce}/\omega_{pe} > 1$, this frequency is approximately equal to the ion plasma frequency ω_{pi}. In discharges in hydrogen and helium, the so-called Buneman frequency, Eq. (III.3), lies near the second harmonic of the ion plasma frequency, $2\omega_{pi}$. To identify the noise spectrum in a plasma in a highly turbulent state is thus a complicated diagnostic problem. In essentially all the experimental work, the decision on whether the turbulent spectrum belongs to a particular instability has been based primarily on the phenomenology of the discharge, on estimated excitation thresholds, and on estimates of the average plasma parameters, rather than on the local values of these parameters. Experimentalists have thus been forced to use various sets of diagnostic methods, of which the most common are the probe, microwave, and spectroscopic methods (the microwave methods work with both the emission from the plasma itself and the scattering of incident microwave signals; the spectroscopic methods make use of the Stark effect and the satellites of forbidden lines of helium).

Harmonics of the ion plasma frequency were detected with a single-turn magnetic probe in the NPR-2 turbulent straight discharge.[113] A magnetic probe was inserted into the turbulent plasma. A loop diameter on the order of 1 cm was chosen for this probe, since it was expected that the nonlinear conversion of ion acoustic waves would result in the excitation of weakly damped whistlers at $\Omega \simeq 2\omega_{pi}$. The signals from the probe were fed directly to the plates of a wide-band oscilloscope.

Figure 95 shows the energy spectra of the turbulent noise for three values of the initial density found from signals 0.1 μs long for a time 0.15-0.2 μs after the beginning of the turbulent heating. As the density decreases, we see a shift of the upper boundary of the spectrum toward lower frequencies; this shift is proportional to $\sqrt{n_e}$. Harmonics are observed near the ion plasma frequency and its second harmonic. The noise intensity is distributed roughly uniformly over the diameter of the current column in the plasma and falls off rapidly outside the plasma.

The energy density of the whistlers was calculated from the condition for their growth in the plasma:

$$\frac{d}{dt}\frac{\tilde{H}^2}{8\pi} \simeq \omega_{pi}\frac{T}{m_e c^2}\frac{E^2}{8\pi}. \tag{VI.6}$$

For an observation time of 0.1 μs, E = 210 V/cm, $n = 10^{13}$ cm^{-3}, and $nT \approx 5\cdot10^{15}$ eV/cm^3, this equation yields $\tilde{H}^2/8\pi \approx 2\cdot10^{-3}$ erg/cm^3, in agreement with the experimental data. The appearance of a noise spectrum near $2\omega_{pi}$ can be ascribed to a nonlinear coalescence of ion acoustic waves

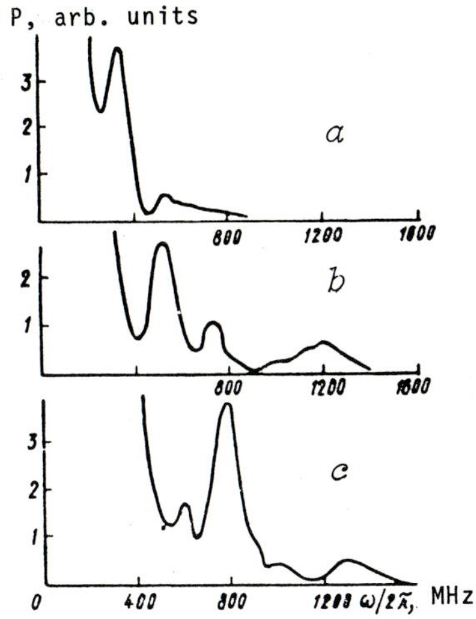

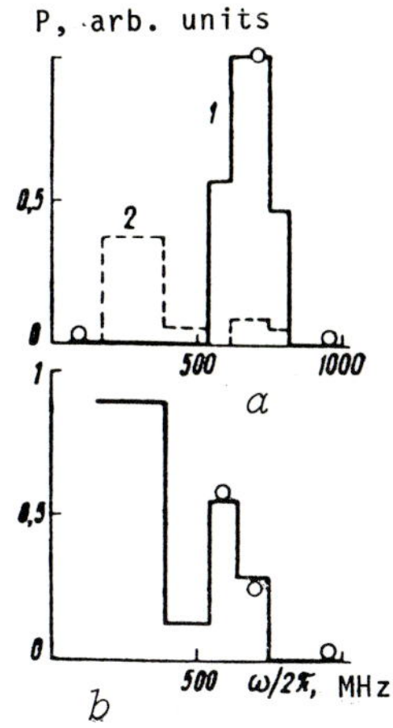

FIGURE 95 Energy spectra of the turbulent noise in the NPR-2 straight hydrogen discharge[113] at various plasma densities: *a*) $2\cdot10^{12}$ cm^{-3}; *b*) $9\cdot10^{12}$ cm^{-3}; *c*) $1.6\cdot10^{13}$ cm^{-3} (H = 7 kOe, E = 240 V/cm, T_e = 500 eV).

FIGURE 96 Spectra of the rf fluctuations in a hydrogen plasma in the VIKhR'-2 device.[142] *a*: Spectrum 20 ns after the beginning of the discharge. *1*) Data from magnetic (solid histogram) and electric (circles) measurements at $n = 6\cdot10^{12}$ cm^{-3}; *2*) the same, at $n = 2\cdot10^{12}$ cm^{-3}. *b*: Spectrum at 80 ns, $n = 6\cdot10^{12}$ cm^{-3} (E = 250 V/cm, H_{max} = 15 kOe, $nT = 10^{15}$ eV/cm^3).

with the frequency ω_{pi} in the plasma.

The time evolution of the turbulence spectra near the ion plasma frequencies was studied in the VIKhR'-2 device.[142] These experiments were carried out because a torus loop electric field $E \approx 250$ V/cm, much stronger than the Dreicer field, excited a purely active current in the plasma, which reached its maximum in $\tau \approx 10\cdot2\pi/\omega_{pi} \approx 20$ ns, i.e., in ten periods of the ion plasma waves. Information on the frequency spectrum was found by several methods: by magnetic and electric probes, from the spectrum of the modulation of the superthermal microwave emission (Fig. 79),

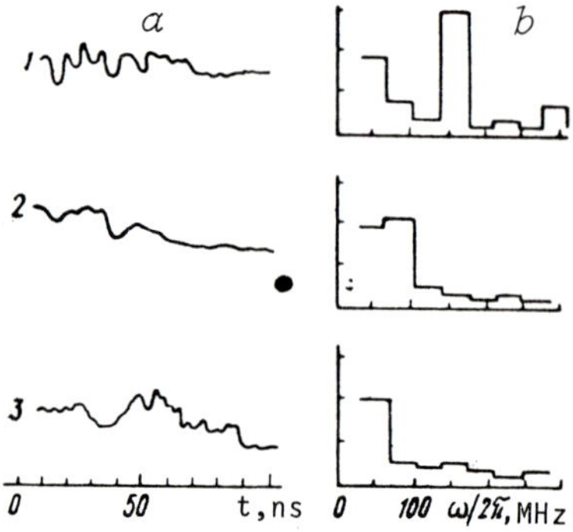

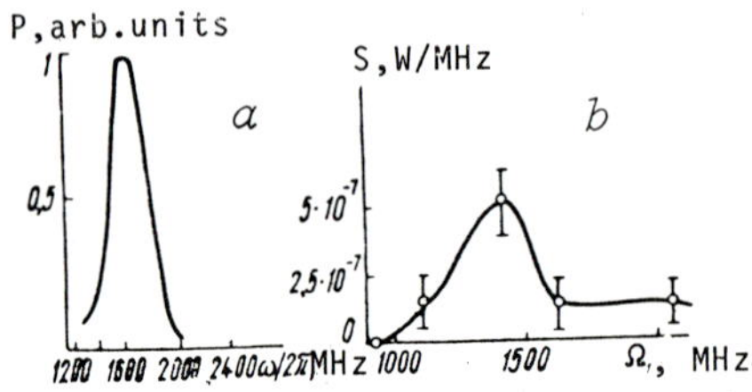

FIGURE 97 Low-frequency waves observed in the plasma in the VIKhR'-2 device with a magnetic probe.[84] *a*: Probe signals. *b*: Voltage spectra. *1*) Xenon plasma, $n = 3\cdot10^{13}$ cm^{-3}; *2*) the same, $n = 1\cdot10^{13}$ cm^{-3}; *3*) hydrogen plasma, $n = 3\cdot10^{13}$ cm^{-3} (E = 400 V/cm, H_{max} = 7 kOe, $nT = 3\cdot10^{16}$ eV/cm^3).

FIGURE 98 Turbulent noise spectra in a hydrogen plasma near $2\omega_{pi}$ in the TOR-4M device.[24,87] *a*) Probe measurements, E = 40 V/cm; *b*) spectral density of the "red" satellite during Raman scattering of a microwave signal at λ_0 = 8 mm (E = 20 V/cm, H = 8 kOe, T_e = 100 eV, $n \simeq 1.7\cdot10^{13}$ cm^{-3}).

and from the Raman scattering of signals with wavelengths λ_0 = 2, 4, and 8 mm (Ref. 103). The spectrum analyzer in the first two of these methods was a fast three-channel analyzer with interchangeable bandpass filters ($\tau \lesssim$ 10 ns).

It was shown that waves with a relatively narrow spectrum near the ion plasma frequency arise at the current front with the shortest measurable delay. The spectrum broadens in $\tau \approx$ 50-60 ns, at both low and high frequencies. The broadening is accompanied by a substantial intensification of the superthermal microwave emission at $\omega \sim \omega_{pe}$ (Fig. 77).

Figure 96 shows the time evolution of the frequency spectrum in the VIKhR'-2. The emergence of a low-frequency tail in this spectrum is accompanied by a lowering of the peak near $\omega \approx \omega_{pi}$. The data from the magnetic and electric probes agree well. The nonelectrostatic waves observed near the frequency $\omega \approx \omega_{pi}$ and its second harmonic, $2\omega_{pi}$, may also be evidence of a nonlinear conversion of ion acoustic waves into whistlers. According to the magnetic measurements and Eq. (VI.6), the relative energy density of the acoustic noise in the VIKhR'-2 was $W_i/nT_i \sim 10^{-2}$.

Experiments in the VIKhR'-2M device[84] on the dependence of the anomalous resistance on the mass of the gas, through the use of a magnetic probe with a loop 4 mm in diameter, yielded information on the low-frequency spectrum, $\omega/2\pi$ = 20-300 MHz, for a fixed plasma density in various gases.

Figure 97 shows some probe signals and the spectra of the low frequency waves. In the xenon plasma there is a frequency shift $\omega \sim \sqrt{n}$ near ω_{pi} and $2\omega_{pi}$. There is also a time delay of 20-25 ns for the low-frequency waves, $\omega \sim$ 20-100 MHz $\ll \omega_{pi}$, in the hydrogen plasma, in agreement with the measurements taken in the VIKhR'-2. These results confirm Rudakov's hypothesis[113] of a nonlinear transformation of the spectrum of short ion acoustic waves excited by the current.

The transformed spectrum of ion acoustic waves was also detected and studied in the TOR-4M device,[24,87] by a contact method (an electrostatic probe) and a contactless method (based on the scattering of a microwave signal at $\lambda_0 \cong$ 8 mm). The spectrum of the signal from the probe was analyzed with a waveguide interference arrangement ($\lambda_0 \cong$ 11-22 cm). The waveguide cross section was chosen such that signals with frequencies on the order of ω_{pi} could not be excited in the waveguide. The noise spectrum near $2\omega_{pi}$ is shown in Fig. 98a.

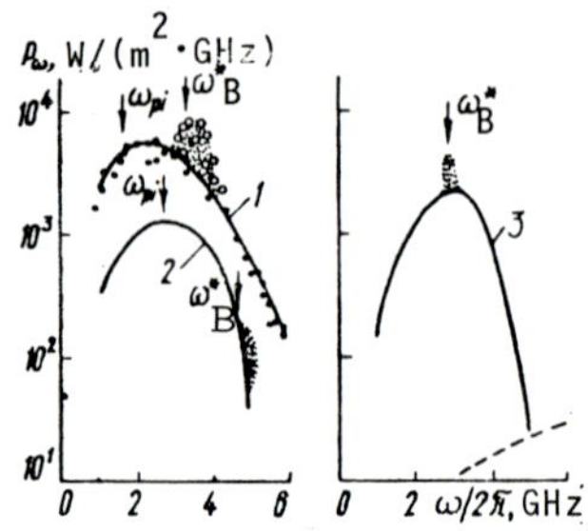

FIGURE 99 Noise spectra of a hydrogen plasma in the TURHE turbulent straight discharge.[271] *1*) E = 400 V/cm, n = $8\cdot 10^{13}$ cm^{-3}; *2*) E = 400 V/cm, n = $1.4\cdot 10^{14}$ cm^{-3}; *3*) E = 250 V/cm, n = $7\cdot 10^{13}$ cm^{-3}.

An interference microwave filter plug (55 dB) was used in experiments on the Raman scattering of microwaves in a turbulent plasma and on detection of the satellite $\omega_0 - \Omega = \omega_0 - 2\omega_{pi}$ in order to suppress the parasitic emission from the high-power microwave source ($P_0 \cong 10^4$ W). The spectrum of the red scattered satellite was analyzed with a two-mirror interference filter with a resolution $\Delta\lambda \approx 10^{-2}$-$4\cdot 10^{-3}$. Figure 98b shows the dependence of the spectral density of the scattered radiation on the Raman frequency shift ($P_{scatt}/P_0 \approx 2\cdot 10^{-8}$). This scattering effect can be explained under the condition $k_0^t \sim k^t \sim k^\Omega << k_D$. In other words, long, high-frequency waves, $\Omega \sim 2\omega_{pi}$, with $k^\Omega << k_D$, must be excited in the plasma. For this dispersion law the plasma may contain whistlers,[118]

$$\Omega = k^\Omega k_z c^2 \frac{\omega_{ce}}{2\pi\omega^2_{pe}},$$

which are of such a nature that the typical wave vector in the plasma,

$$k^\Omega \approx \frac{\omega_{pe}}{c}\left(\frac{\omega_{pi}}{\omega_{ce}}\right)^{1/2},$$

may in fact satisfy the scattering condition, $k^t = k_0^t + k^\Omega$. The whistler field amplitudes found experimentally from the

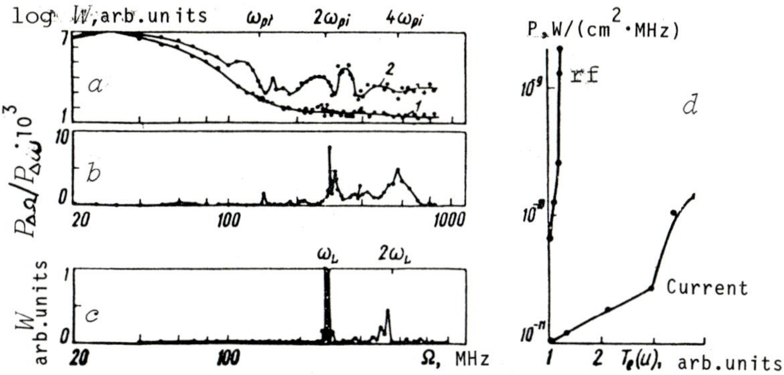

FIGURE 100 *a,b,c*) Turbulent noise spectra in the URAGAN stellarator[52,263,132,80]; *d*) intensity of the noise measured outside the plasma vs the electron temperature T_e. Current discharge in helium: *a* -- Noise power spectrum according to magnetic measurements. *1*) Steady-state spectrum; *2*) sporadic spectrum. *b* -- Modulation spectrum of the sporadic microwave emission in the ω_{pe} ($\Delta\Omega/\Delta\omega \approx 10^{-3}$, $\Delta\omega \approx 0.1\omega_{pe}$) region. Radio-frequency discharge in deuterium: *c* -- Magnetic measurements.

probe and microwave measurements in the TOR-4M device are $H \approx$ 1-1.3 Oe.

Contactless methods have been used to study the turbulence spectra in hot plasmas in several experiments, e.g., in the TURHE device[265,271] and the URAGAN device.[132,264] External magnetic antennas detected intense "emission" into vacuum.

In the TURHE experiments the emission was detected with two wide-band helical antennas, one used as a monitor. The spectrum is analyzed with a bandpass filter with $\Delta\omega \approx 35$ MHz. The structure in the noise spectrum near the ion plasma frequencies is linked with the appearance of sporadic bursts near the frequency $\omega^*_{BB} \approx 2\omega_{pi}$ [see Eq. (III.3)]. Figure 99 shows the emission spectra from the TURHE device. Against the background of a steady-state spectrum there are some intense, time-varying pulsed

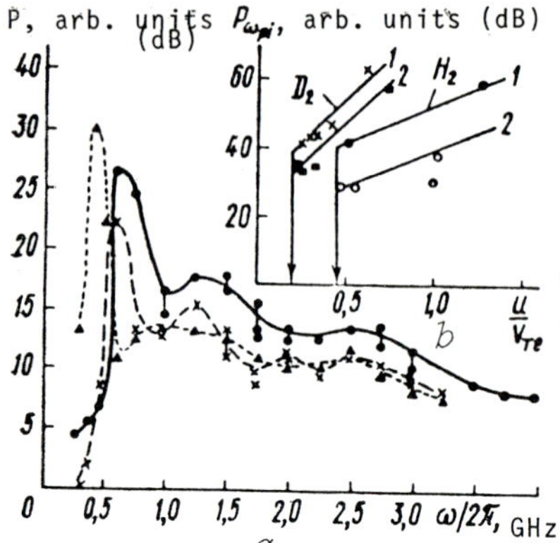

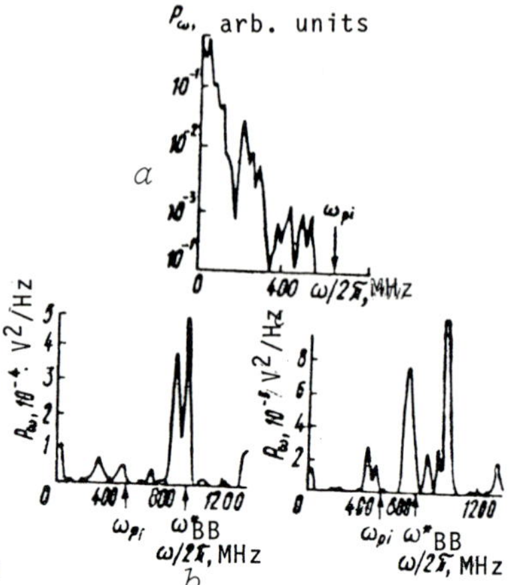

FIGURE 101 *a*: Time evolution of the turbulence spectra in the ALCATOR tokamak.[276] *b*) Threshold in the dependence of the excitation of the ion plasma noise at ω_{pi} on the ratio of the current velocity to the thermal velocity for various impurity parameters in deuterium and hydrogen. H = 40 kOe, E = (3-1)$\cdot 10^{-3}$ V/cm, $T_e \approx T_i \simeq 1$ keV, I = 100 kA; ●) t_1 = 20 ms; x) t_2 = 40 ms; ▲) t_3 = 150 ms; $n_1 = 9\cdot 10^{12}$ cm^{-3}, $n_2 = 8\cdot 10^{12}$ cm^{-3}, $n_3 = 5\cdot 10^{12}$ cm^{-3}.

FIGURE 102 Turbulence spectra detected with a double electrostatic probe in the hydrogen plasma in the TWIST stellarator.[225] *a*) $n = 10^{13}$ cm^{-3}, E = 200 V/cm, u = 1.1x10^8 cm/s; *b*) $n = 5\cdot 10^{12}$ cm^{-3}, E = 450 V/cm, u = (2.4-3.2)$\cdot 10^8$ cm/s, t = 110-140 ns; *c*) the same, for t = 190-220 ns, $u = 3.5\cdot 10^8$ cm/s. H = 3 kOe.

spectra.

A sporadic excitation of noise near harmonics of the ion plasma frequency was observed in the URAGAN stellarator. A comparative study was made in those experiments of the emission of noise near the ion plasma frequencies and of the conditions for the excitation of this noise for various temperature models: $T_e \gg T_i$ for turbulent heating by a current and $T_i \gg T_e$ for rf ion cyclotron heating ($\omega_{ci} \approx 10$ MHZ $\ll \omega_{pi}$). Signals were detected by loops

110 cm in diameter outside the plasma and analyzed by a four-channel system of superheterodyne receivers, which were subjected to a special calibration. The spectra in the discharges of the various types were compared for constant values of the plasma density, of the magnetic field ($\omega_{ce}/\omega_{pe} > 1$), of the gas pressure $[(nT_e)_{curr} \cong (n_i T_i)_{rf}]$, and of the power pumped into the plasma ($W^0_{curr} \approx W^0_{rf}$).

Figure 100 shows frequency spectra of the superthermal turbulence in the discharge with a current and in the rf discharge, along with the dependence of the power density at the magnetic loop antenna on the electron temperature T_e. We see that a steady-state spectrum is excited in the discharge with a current (spectrum *1* in Fig. 100a) which is similar in nature to the steady-state ion acoustic spectrum reported by Kadomtsev,[107] and there is also an explosive sporadic spectrum (spectrum *2* in Fig. 100) near the ion plasma frequency and its harmonics.

The appearance of intense, time-varying noise bursts at $\Omega \sim 2\omega_{pi}$ in the discharge with a current was accompanied by the emission of superthermal x radiation with energies from 1 to 10 keV from a target in the plasma. In addition, there was sporadic superthermal microwave emission, for which the modulation spectrum is shown in Fig. 100b (see also Fig. 92). During rf heating in an ion-hot plasma, $T_i >> T_e$, there is an intense, steady-state superthermal spectrum with a narrow peak near the lower hybrid frequency and a broader peak near its second harmonic (Fig. 100c).

Proof that different mechanisms are operating in the discharge with a current and in the rf discharge to excite the turbulent noise near the lower hybrid resonance and its second harmonic, $\Omega \sim 2\omega_{pi}$, comes from the dependence of the intensity of this noise on the electron temperature (Fig. 100d). It can be seen from this figure that during ion cyclotron heating the intensity of the approximately steady-state superthermal noise is highly sensitive to the electron temperature, while in the current discharge the relationship is weaker. For equal pressures, $(n_e T_e)_{curr} \approx (n_i T_i)_{rf}$, and for equal values of the power pumped into the plasma, $W^0_{curr} \approx W^0_{rf}$, the peak intensity of the noise "emitted" by the discharge with a current ($T_i \ll T_e$) is much lower than that of the noise from the rf discharge ($T_e \ll T_i$). The difference may be due to a trapping of the noise in the electron-hot plasma.

The results from the URAGAN stellarator showed that different plasma-heating methods led to the same type of superthermal turbulent noise, which was ultimately due to the excitation of waves near the lower hybrid resonance. This superthermal electromagnetic noise might have resulted from a) an ion acoustic instability[147] driven by the longitudinal current in a plasma with $E/E_{Dr} > 1$, $T_e \gg T_i$, and a drift velocity stabilized at $u/v_S \lessapprox 10$; or b) an electron acoustic instability,[151] driven by a transverse current, $\frac{u_e^{\perp}}{v_{Ti}} \gtrsim 1$, in the plasma of an rf discharge, with $T_i \gg T_e$. In turn, the linear and nonlinear conversion of small-wave-number sound near the lower hybrid resonance could have led to the appearance of long-wave turbulent noise, with $k \ll k_D$, with a frequency spectrum $\Omega \approx \omega_L \simeq \omega_{pi}$ and $\Omega \sim 2\omega_{pi} \sim 2\omega_L$. A small-wave-number ion acoustic wave source of this type might possibly be the primary reason for the sporadic "emission" of noise at

ω_{pi} and $2\omega_{pi}$ and of superthermal microwave emission at $\omega \sim \omega_{pe}$ in a plasma with hot electrons, $T_e >> T_i$. This effect probably results from a modulational instability and from the growth of waves near the lower hybrid resonance, which highly perturb the plasma density and form cavitons. Analysis of the modulational spectrum of the superthermal microwave emission, $\omega^t = \omega^l_{pe} + 2\omega_{pi}$ (Figs. 92 and 100b), shows that there is a strong, nonlinear interaction of electron plasma waves and ion plasma waves in the plasma. For the plasma parameters in the URAGAN stellarator during the current heating the wave numbers of the low-frequency turbulent noise are $k^\Omega_\parallel \sim k^l_\parallel \sim k^t \approx 2.5\ \text{cm}^{-1} << k_D = 150\ \text{cm}^{-1}$. In this case the longitudinal phase velocity of the low-frequency waves, $v^\parallel_\varphi = \frac{\omega_L}{k^\Omega_\parallel} \lessapprox 3.6 \cdot 10^8$ cm/s, is lower than the electron thermal velocity, $v_{Te} \approx 5 \cdot 10^8$ cm/s, and this circumstance may be responsible for the trapping and growth of these waves in the plasma.

In the rf discharge, only thermal microwave noise is observed near the upper hybrid frequency, $\omega_h^2 = \omega_{pe}^2 + \omega_{ce}^2$ with a fine structure which indicates Raman scattering of electromagnetic radiation by density fluctuations, $\Omega \sim \omega_L$, in accordance with the conservation laws $\omega^{t'} = \omega_h^t + \Omega$ and $\mathbf{k}^{t'} = \mathbf{k}^t + \mathbf{k}^\Omega$. Since $\mathbf{k}^{t'} \cong \mathbf{k}^t \sim \mathbf{k}^\Omega$, the phase velocity of a low-frequency wave, $v^\parallel_\varphi = \frac{\omega_L}{k^\Omega_\parallel} > 4 \cdot 10^8$ cm/s, is higher than the electron thermal velocity, $v_{Te} \simeq 2.5 \cdot 10^8$ cm/s, and approximately equal to the phase velocity of an electron acoustic wave, $v^\parallel_\varphi \simeq v^e_s = \left(\frac{T_i}{m_e}\right)^{1/2}$, in an ion-hot plasma ($T_i = 300$ eV, $T_e = 30$ eV). This fact may be responsible for the effective "emission" of waves, through a linear conversion from the region of the lower hybrid resonance in an inhomogeneous plasma (Golant and Piliya[64]).

In experiments in strong magnetic fields, $\omega_{ce}/\omega_{pe} > 1$, in the ALCATOR tokamak,[276] with a drift velocity stabilized at $u = 0.2v_{Te}$ in deuterium and at $u = 0.45v_{Te}$ in hydrogen, an intense superthermal noise near the lower hybrid resonance was detected with an electrostatic probe in the shadow of the limiter. Figure 101 shows spectra of this noise at various times. There are intense peaks near $\omega_L \approx \omega_{pi}$ and higher harmonics. Also shown in this figure is the behavior of the noise intensity in the discharge as a function of the ratio u/v_{Te}. Here we can see that there is a threshold for the excitation of the high-frequency turbulence in various gases with various levels of contamination by impurities.

Turbulent noise with $\omega \approx \omega_{pi}$ was excited a few milliseconds after the synchrotron radiation, $\omega \sim \omega_{ce}$, and the hard x radiation. The appearance of this noise was always correlated with the emission of fast neutral atoms, indicating a possible relationship between the anomalous heating of ions and the observed turbulence.

Schneller et al.[276] believe that the typical spectral shape, with a sharp decay in the noise intensity at $\omega < \omega_{pi}$ (Fig. 101), which was observed in the ALCATOR at $\frac{\omega_{ce}^2}{\omega_{pe}^2} \gg 1$ and in the presence of a large number of runaway electrons, is not evidence of the classical two-stream ion acoustic instability. As has already been mentioned (Section 2 of Chapter VI), it is possible that low-frequency plasma-wave modes, $\omega = \frac{k_\parallel}{k}\omega_{pe}$, may be excited under the conditions corresponding to the plateau on the distribution function of the accelerated electrons; for the parameters of the ALCATOR, the spectrum of these waves stretches up to $\omega/2\pi > 1$ GHz (Ref. 204).

Much information on the low-frequency turbulence spectrum was obtained by Hamberger, Jancarik et al.[225] and by Sharp and Mrowka[278] in the TWIST stellarator in measurements with an electrostatic probe and by microwave scattering. The investigators used a high-impedance, wide-band ($\Delta\omega \simeq$ 3-1000 MHz), double electrostatic probe (the distance between the pins was 1 mm; their length was 1 mm; and their diameter was 0.1 mm), intending to observe electrostatic ion acoustic waves.

The electric field in the TWIST experiments was exceptionally high, E = 400 V/cm. Three characteristic regimes (A, B, and C) were identified at different values of the drift velocity u. At $u < 2\cdot 10^8$ cm/s the spectrum was at frequencies $\omega < \omega_{pi}$, and the fluctuations were oriented parallel to the current (regime A). The perpendicular component increased with increasing u, at $u \gtrsim 2\cdot 10^8$ cm/s. The nature of the fluctuations changed radically at the threshold values $u \approx 2\text{-}3\cdot 10^8$ cm/s (regime B). The spectrum in hydrogen discharges was concentrated near the frequency $\omega^*_{BB} \sim 2\omega_{pi}$, while that in argon was near $\omega^*_{BB} \sim 3\omega_{pi}$. In regime C, at very low densities, $n \simeq 2\cdot 10^{11}$ cm^{-3}, and at fields $E/E_{Dr} \approx 4000$, intense sporadic bursts at $\omega_{pe}/4$, $\omega_{pe}/2$, and ω_{pe} were detected.

Figure 102 shows some typical turbulence spectra in regimes A and B in hydrogen discharges. Hamberger and Friedman were the first to point out that in a dense plasma with $n \approx 10^{13}$ cm^{-3}, at relatively low current velocities,

$u \gtrsim 2 \times 10^8$ cm/s, in regime A, the power spectrum of the fluctuations which is measured has a distribution like Kadomtsev's steady-state ion acoustic spectrum.[107] This spectrum is shown in Fig. 103.

Microwave scattering (at $\lambda_0 = 2.4$ mm and 8.8 mm) was observed in the plane of the torus in the TWIST stellarator[278] at various angles $\vartheta = \mathbf{k}_p \wedge \mathbf{I}$ between the current (**I**) and the wave vector (**k**) of the fluctuations in the plasma. These measurements were carried out by arranging oblique incidence of the wave vector $\mathbf{k}_0$ of the external microwave signal with respect to the current **I** and by varying the scattering angle in the plane of the torus.

Figure 104 shows fluctuation spectra in the scattered signal at $\lambda_0 \approx 2.4$ mm obtained through a homodyne conversion with various analysis methods (bandpass filters and Fourier analysis). Discharge conditions were arranged in the TWIST such that the drift velocity was stabilized at $u \approx (4\text{-}5)v_s$, and conditions were arranged for occurrence of the ion acoustic instability in a hydrogen plasma. The value $u \approx 7 \cdot 10^8$ cm/s, we might note, corresponded to regimes B and C as defined above. It can be seen from this figure that the noise spectrum stretches far beyond the frequencies ω_{pi} and ω^*_{BB}. Long-wave noise with $\mathbf{k}_p \perp \mathbf{I}$, where $\mathbf{k}_p \sim \mathbf{k}_0$, is excited. According to Raman data, the turbulence spectrum in the case $\vartheta = 90°$ evolves over a time of 250 ns from low frequencies, 100 MHz, to 700 MHz. During the same time interval, intense superthermal microwave emission with $T_{Br} \simeq 10^5 T_e$ is observed

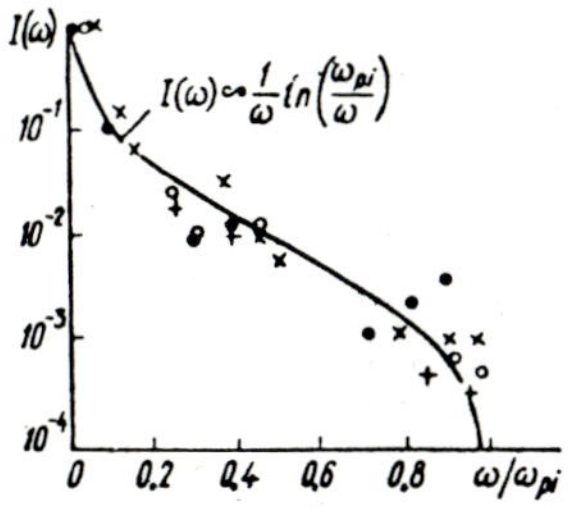

FIGURE 103 Spectrum of the fluctuation power density under ion-acoustic-turbulence conditions in the TWIST.[225] The experimental data are shown for four discharges under identical conditions. Curve) Spectrum reported by Kadomtsev[107]; points) experimental maximum values of $\langle V^2(\omega)\rangle$ for $n = 10^{13}$ cm^{-3}, $u = 1.1\cdot10^8$ cm/s.

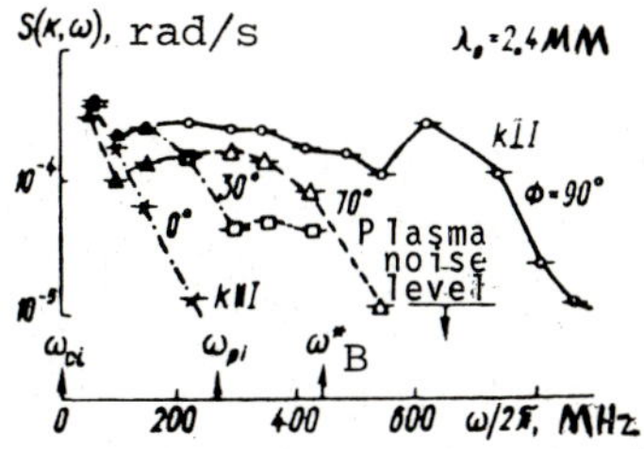

FIGURE 104 Spectrum of the fluctuations in a scattered microwave signal (λ_0 = 2.4 mm) for various orientations of the wave vector **k** with respect to the plasma current in the TWIST.[278] t = 250 ns, E = 500 V/cm, $n = 2\times10^{12}$ cm^{-3}, H = 3 kOe, $T_e \simeq 10^4$ eV, I = 4 kA.

(Fig. 82). Sharp and Mrowka[278] believe that this microwave emission may indicate a relationship between the high-frequency density fluctuations and an electromagnetic instability.

Methods involving Raman microwave scattering in plasmas have also been used widely for studies of turbulence in several other devices. As we have already mentioned, for example, the set of wavelengths λ_0 = 2, 4, 8, and 30 mm was used simultaneously for a study of low-frequency turbulence fluctuations at $\omega < \omega_{pi}$ in the VIKhR'-2 device.[103,142] It was found that the long-wave ion acoustic waves in the plasma are highly anisotropic, and their linear dispersion was observed.

An extensive series of studies on microwave scattering in turbulent plasmas was carried out in the high-current straight discharge in the THM device[301] at the wavelength λ_0 = 4 mm ($n = 6\text{x}10^{12}$ cm^{-3}, E = 220 V/cm, T_e = 1 keV, T_i = 500 eV, H = 2.2 kOe). An interesting aspect of the scattered signal was a peak near the ion plasma frequency, $\omega_{pi}/2\pi \approx$ 500 MHz; this peak existed only during the first 0.8 μs of the heating.

The time evolution of the noise spectrum in the turbulent linear discharge in the BSG II device[184] was studied with an electrostatic probe and by microwave scattering. Intense peaks were observed near ω_{pi} and its harmonic $2\omega_{pi}$. In these experiments, as in the VIKhR'-2 experiments,[103,142] the deformation of the spectrum was accompanied by the gradual growth of a low-frequency tail and by a shrinkage of the peaks at $\omega \sim \omega_{pi}$ and $\omega \sim 2\omega_{pi}$.

The most important result on the dynamics of turbulent processes was obtained in the MACH II high-current straight discharge (Adati et al.[185]): In the highly turbulent stage of the discharge, a flattening of the electron distribution function at low energies, $T_e \approx$ 10-20 eV, was observed, in association with noise at $\omega \lesssim \omega_{pi}$. The electron velocity distribution was studied by Thomson scattering with an eight-channel system having a high spatial and temporal resolution. In a hydrogen discharge with $n = 5 \cdot 10^{12}$ cm^{-3}, T_e = 50 eV, H = 7.5 kOe, I = 5 kA, and $u = 0.5v_{Te}$ the observed flattening of the distribution function can be ascribed to trapping of electrons in the potential wells of ion acoustic waves.[194,196] The time required for this flattening to occur is on the order of $100/\omega_{pi}$; no saturation of the instability was observed.

Raman scattering of a microwave signal at λ_0 = 6.2 mm in this apparatus, with the scattered signal observed at an angle $\vartheta = 90°$, revealed a narrow-band spectrum, $\Delta\omega$ = 10-20 MHz, which shifted along the frequency scale over time in accordance with the increase in the electron temperature in the turbulent discharge: Δt = 1-1.2 μs, Ω = 20-80 MHz, and T_e = 100-1000 eV (according to laser scattering and x-ray measurements; Fig. 30).[188] The Raman-scattering results obtained from the MACH II are in complete agreement with results obtained earlier in the VIKhR'-2 turbulent discharge,[103,142] where the spectrum of the scattered microwave signal at λ_0 = 8 mm also revealed a narrow peak with a frequency $\Omega \sim \sqrt{T_e}$ (Ω = 25-110 MHz at E = 50-250 V/cm). Furthermore, microwave-scattering experiments[254] in the MACH II also revealed a linear dependence $\Omega(k)$, corresponding to the dispersion for long-wave ion sound, as in the VIKhR'-2 device.[142,103]

These experimental facts are direct proof that an ion acoustic instability with a normal dispersion law, $\omega_s = k_s v_s$, is driven in the turbulent current discharge. These facts are furthermore in satisfactory agreement with the theory for the Raman scattering of microwaves by ion acoustic waves.[102,150]

Direct experiments to determine the dispersion laws for

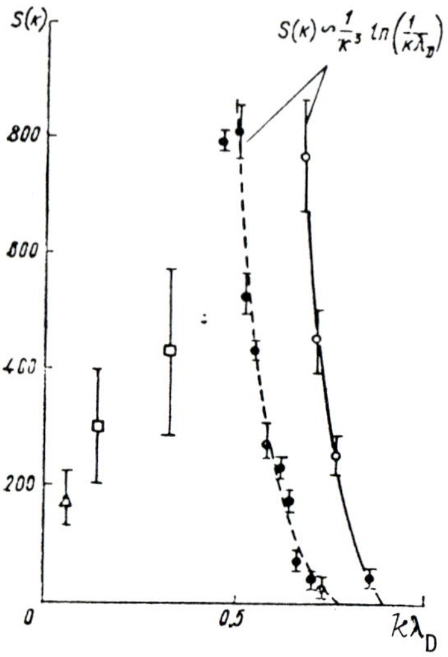

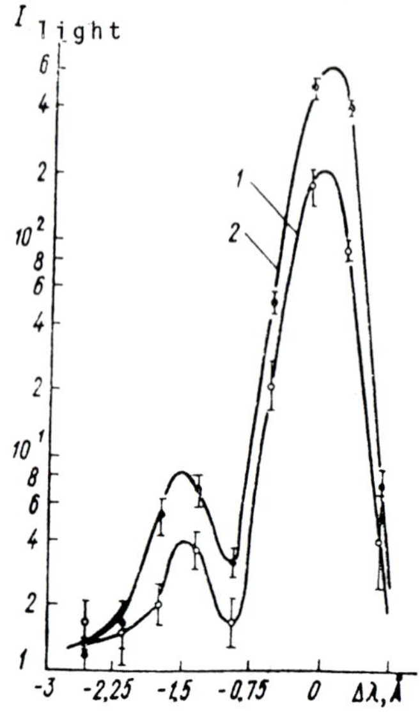

FIGURE 105 Raman spectrum of the short-wave turbulence in the plasmas in the TARANTULA I and TARANTULA II devices[258] according to the scattering of beams from CO_2 (λ_0 = 10.6 µm) and ruby (λ_0 = 694 nm) lasers. o) TARANTULA I, 694 nm, $\lambda_D = 1.1 \cdot 10^{-6}$ m; ●) TARANTULA I, 694 nm, $\lambda_D = 7.4 \cdot 10^{-7}$ m; Δ) TARANTULA I, 10.6 µm, $\lambda_D = 1.1 \cdot 10^{-6}$ m; □) TARANTULA II, 10.6 µm, $\lambda_D = 7.4 \cdot 10^{-7}$ m. Shock waves: z pinch (TARANTULA I), T_e^0 = 22 eV, $n = (6.4\text{-}10.5) \cdot 10^{14}$ cm^{-3}, $M = 2.7$, H = 1.2 kOe; z pinch (TARANTULA II), T_e^0 = 18 eV, $n = 9.6 \cdot 10^{14}$ cm^{-3}, H = 1.2 kOe, I = 800 kA, t = 0.7 µs.

FIGURE 106 Emission spectrum near the line HeI 4921.93 Å in the SIRIUS stellarator[28] at various times. *1*) t = 25 µs, $n = 4.5 \cdot 10^{12}$ cm^{-3}; *2*) t = 50 µs, $n = 1.9 \cdot 10^{13}$ cm^{-3} (H = 16 kOe, E = 0.7 V/cm, $T_e \simeq$ 200 eV).

small-wave-number ion acoustic waves were carried out in the TARANTULA I and TARANTULA II devices,[258] in a high-current, large-diameter z pinch in a magnetic field. Small-angle scattering of the beams from ruby and CO_2 lasers was used for diagnostics, in accordance with Eq. (VI.1).

Figure 105 shows the small-wave-number fluctuation spectrum in the TARANTULA I and the TARANTULA II. It was observed that the peak intensity of the ion acoustic noise is at $k\lambda_D = 0.5$. The spectral shape, $S(k) \sim \frac{1}{k^3} \ln \frac{1}{k\lambda_D}$, corresponds to the theoretical spectrum derived by Kadomtsev.[109]

Spectroscopic diagnostic methods can yield quantitative results on the turbulence spectrum.[182,220,36] The positions of the satellites on the wings of the hydrogen lines can be used to determine the temporal and spatial integrals of the fields in the ion acoustic spectrum.

In Zavoĭskiĭ's experiments[93] in the NPR-2 there was a pronounced broadening of the $H\beta$ and the $H\alpha$ lines (~1.8 Å and ~3.6 Å, respectively) during the first stage of the current flow, for about 0.4 μs (Fig. 74). Observation of the lines through a polarizer and measurement of the line profiles in two mutually perpendicular orientations of the polarizer made it possible to eliminate all Doppler- and Holzmark-broadening effects. The electric field of the ion acoustic fluctuations was estimated to be $E_S \approx$ 15-20 kV/cm. A rough measure of the asymmetry of the electric field distribution is the ratio of the distances between the positive and negative peaks in the difference profile from the two components (σ and π). This type of anisotropy of ion acoustic waves was observed in the NPR-2 device.[93,182] The degree of turbulence estimated from the Stark effect in the TOR-4M, for example, was $W_S/nT = 2 \cdot 10^{-3}$. The extreme electric field of the turbulent microfluctuations in the TURHE device[245,273] was 50 kV/cm.

The turbulence spectra in the SIRIUS stellarator were studied with an electrostatic probe[30] and also on the basis of the spectrum of the satellites of forbidden lines of helium, HeI 4921.93 Å. Figure 106 shows emission spectra from the SIRIUS[28] near the HeI line at various times. These spectra were obtained with a ten-channel photoelectric spectrometer. We see an intensity peak in the turbulent discharge near the position of the forbidden line. The electric field in the plasma in the SIRIUS stellarator was 2 kV/cm at a degree of turbulence $W_s/nT = 3\cdot10^{-4}$–$2.5\cdot10^{-3}$.

Several important facts have been established in the extensive experimental work reviewed here on the spectra and dynamics of low-frequency turbulence in plasmas with currents.

1. When the drift velocity is stabilized at a level $u \tilde{>} v_s$, low-frequency, approximately steady-state spectra are excited with an approximately linear dispersion law $\omega_s = k_s v_s$. The spectral density distributions of the fluctuations observed experimentally are $S(k) \sim k^{-3} ln(k\lambda_D)^{-1}$ and $I(\omega) \sim \omega^{-1} ln(\omega_{pi}\omega^{-1})$ in the long-wave region $(k < k_D)$ and are similar to the steady-state ion acoustic spectra predicted by Kadomtsev's theory.[109]

2. In the highly turbulent stage of current discharges, with $u = 10v_s$, the steady-state spectra just described are accompanied by relatively narrow-band, time-varying spectra at the ion plasma frequency and its harmonics, which are of a nonelectrostatic nature. As a rule, intense superthermal sporadic microwave emission also occurs under these conditions.

3. The high level of approximately steady-state turbulent noise, $W_s/nT \sim 2\cdot10^{-4}$–$2\cdot10^{-3}$, which has been found by spectroscopic diagnostic methods in several experiments, completely explains the approximately steady-state anomalous resistance and the turbulent heating of plasmas.

The time-varying spectra correspond to a higher degree of turbulence, $W_i/nT \sim 10^{-2}$. Under these conditions the strong-turbulence state in a current-carrying plasma is somewhat reminiscent of the state and phase-transition dynamics of a boiling liquid.

CONCLUSION

In this monograph we have been able to do no more than draw a general picture of the present state of experimental research on the anomalous resistance of plasmas and their turbulent heating by electric currents. We have appealed to the theory on turbulent plasmas as little as possible, solely when necessary to reach an understanding of the experiments. All questions related to the anomalous resistance and turbulent heating at the fronts of collisionless shock waves have been omitted; the reader interested in pursuing these questions is referred to the monographs by Kadomtsev, Sagdeev, Rudakov, Ryutov, Vedenov, et al.

The physics of turbulent heating has been under study for the past 20 years in more than 40 experimental devices at 15 of the world's leading laboratories. This research has shown that turbulent current heating is an extremely complicated physical phenomenon, which changes in nature at certain critical electric and magnetic fields and which depends on the particular type of energy loss and the particular properties of the device used to confine the hot plasma.

We have mentioned four experimental regimes, differing in the strengths of the electric and magnetic fields; for these regimes, however, stabilization of the electron drift velocity is an obligatory condition.

The adoption of the single, physically clear assumption that the electron drift velocity stabilizes at the level required to drive an instability in a current-carrying plasma has proved completely adequate for practical calculations for most experiments.

There remains some uncertainty regarding the events responsible for the plateau on the $\sigma(E)$ dependence in strong electric fields. More experiments are required on the effect of the mass of the working gas and of the magnetic field on the anomalous resistance of current-carrying plasmas. The mechanism for the turbulent heating of ions requires detailed study. The effective heating of ions to temperatures approaching the electron temperature cannot be explained on the basis of classical ion sound.

Methods for studying the dynamics of ion and Langmuir turbulence in plasmas with currents require further development. The use of turbulent heating in large systems, where a skin effect in the current flow will play a major role, requires further study.

It should be noted that the physical effect which underlies the turbulent-heating method — the anomalous electrical conductivity of a current-carrying plasma — is seen frequently in a wide variety of natural and laboratory situations, e.g., in laser heating of plasmas, upon the onset of the disruptive instability in large tokamaks, in certain stellarator experiments, in Marshall guns (used to produce high-energy plasma streams), and in the interaction of relativistic beams with plasmas. Several natural phenomena, e.g., the solar radio emission, may also be explained as manifestations of microinstabilities in plasmas.

The large number of situations which exhibit the effects of collective interactions in current-carrying plasmas justifies the major research effort on this question.

(Note added by the authors after the revision)

Since the Russian version of the book saw the light a number of important works have been accomplished that significantly enriched our knowledge of the collective phenomena in a current carrying plasma.

Bychenkov and Silin have carried out an in-depth theoretical analysis of the ion-sound instability which made room for explication of the experimentally observable spectrum and turbulent pulsation level and disposed of the earlier disagreement among various theoretical models.*

In experiments on the TRIAM-1, Uragan-2, Alcator and Tortur II facilities the ion turbulent heating has been more thoroughly investigated under the conditions of magnetized plasma at a high electric field voltage. Under the same conditions on the Uragan-2 stellarator the free acceleration breakaway effect in an MHD stable turbulent discharge has been achieved.**

---

*V. Yu. Bychenkov and V. P. Silin, "Plasma ion sound turbulence" [in Russian], *Zh. Eksp. Teor. Fiz. 82*, 1886-1903 (1982).

**N. F. Perepelkin, V. A. Suprunenko, A. S. Slavny, M. P. Vassilyev, A. G. Dikii, V. D. Kotsubanov, B. V. Kravchin, and A. E. Kulaga, in: "The dynamics of microprocesses and current induced turbulent heating of ions on the Uragan-2 stellarator," Res. Rep. IPPJ-548, Oct. 1981, Nagoya, Japan. Preprint KhFTI 81-17.

REFERENCES

1. Adamov, I. Yu., L. A. Dushin, V. I. Kononenko, and O. S. Pavlichenko, *At. Energ. 16*, 99-105 (1964).
2. Al'bikov, Z. A., M. I. Ivanov, V. I. Kozacha, O. V. Kozlov, V. D. Rusanov, and S. S. Sobolev, *Zh. Tekh. Fiz. 44*, 2421-2423 (1974) [*Sov. Phys. Tech. Phys. 19*, 1498 (1975)].
3. Alfvén, H., and K. H. Fälthammer, *Cosmical Electrodynamics*, Oxford Univ. Press, London (1963).
4. Antipov, S. V., M. V. Nezlin, E. N. Snezhkin, and A. S. Trubnikov, Preprint IAÉ-2784, Kurchatov Institute of Atomic Energy, Moscow, 1977.
5. Antonov, A. S., O. A. Zinov'ev, V. D. Rusanov, and A. V. Titov, *Zh. Eksp. Teor. Fiz. 58*, 1567-1569 (1970) [*Sov. Phys. JETP 31*, 838 (1970)].
6. Aranchuk, L. E., E. K. Zavoĭskiĭ, D. N. Lin, and L. I. Rudakov, *Pis'ma Zh. Eksp. Teor. Fiz. 15*, 33-36 (1972) [*JETP Lett. 15*, 22 (1972)].
7. Artsimovich, L. A., *Controlled Thermonuclear Reactions*, Gordon and Breach, New York (1968).
8. Artsimovich, L. A., *Elementary Plasma Physics* [in Russian], Gosatomizdat, Moscow (1969).
9. Artsimovich, L. A., *Nucl. Fusion 12*, 215-219 (1972).
10. Afrosimov, V. V., V. A. Glukhikh, V. E. Golant, A. N. Zaĭdel', E. G. Komar, B. P. Konstantinov, G. M. Malyshev, I. F. Malyshev, N. A. Monoszon, A. M. Stolov, and N. V. Fedorenko, *Zh. Tekh. Fiz. 30*, 1381-1393 (1960) [*Sov. Phys. Tech. Phys. 5*, 1311 (1961)].
11. Afrosimov, V. V., I. P. Gladkovskiĭ, Yu. S. Gordeev, I. F. Kalinkevich, M. P. Petrov, and N. V. Fedorenko, *Zh. Tekh. Fiz. 30*, 1469-1484 (1960) [*Sov. Phys. Tech. Phys. 5*, 1389 (1961)].
12. Akhiezer, A. I., and Ya. B. Faĭnberg, *Dokl. Akad. Nauk SSSR 69*, 555-563 (1949).
13. Akhiezer, A. I., and Ya. B. Faĭnberg, *Zh. Eksp. Teor. Fiz. 21*, 1262-1269 (1951).
14. Akhiezer, A. I., N. G. Prokhoda, and A. G. Sitenko, *Zh. Eksp. Teor. Fiz. 33*, 750-757 (1957) [*Sov. Phys. JETP 6*, 576 (1958)].
15. Akhiezer, A. I., G. Ya. Lyubarskiĭ, and R. V. Polovin, *Zh. Eksp. Teor. Fiz. 40*, 963-969 (1961) [*Sov. Phys. JETP 6*, 673 (1961)].
16. Akhiezer, A. I., I. A. Daneliya, and N. L. Tsintsadze, *Zh. Eksp. Teor. Fiz. 46*, 300-306 (1964) [*Sov. Phys. JETP 19*, 208 (1964)].
17. Babykin, M. V., E. K. Zavoĭskiĭ, L. I. Rudakov, and

V. A. Skoryupin, *Nucl. Fusion,* Supplement *3*, 1073-1077 (1962).

18. Babykin, M. V., P. P. Gavrin, E. K. Zavoǐskiǐ, and L. I. Rudakov, *Zh. Eksp. Teor. Fiz. 47*, 1597-1600 (1964) [*Sov. Phys. JETP 20*, 1073 (1965)].
19. Babykin, M. V., P. P. Gavrin, E. K. Zavoǐskiǐ, L. I. Rudakov, and V. A. Skoryupin, in: *Plasma Physics and Controlled Nuclear Fusion, 2*, (IAEA) Unipub, New York (1966), pp. 851-862.
20. Babykin, M. V., P. P. Gavrin, E. K. Zavoǐskiǐ, S. P. Nedoseev, L. I. Rudakov, and V. A. Skoryupin, *Zh. Eksp. Teor. Fiz. 52*, 643-656 (1967) [*Sov. Phys. JETP 25*, 421 (1967)].
21. Babykin, M. V. and G. E. Smolkin, in: *Plasma Physics and Controlled Nuclear Fusion, 1*, (IAEA) Unipub, New York (1969), pp. 129-142.
22. Babykin, M. V., E. D. Volkov, P. P. Gavrin, B. A. Demidov, E. K. Zavoǐskiǐ, L. I. Rudakov, V. A. Skoryupin, V. A. Suprunenko, E. A. Sukhomlin, Ya. B. Faǐnberg, and S. D. Fanchenko, *Otkrytiya, Izobreteniya, Prom. Obraztsy, Tovar. Znaki* (No. 32) (1972).
23. Balakhanov, V. Ya., V. K. Zhivotov, O. A. Zinov'ev, G. D. Myl'nikov, V. D. Rusanov, and A. V. Titov, *Zh. Eksp. Teor. Fiz. 56*, 439-443 (1969) [*Sov. Phys. JETP 29*, 240 (1969)].
24. Balakhanov, V. Ya., and V. K. Zhivotov, *Izv. Vyssh. Uchebn. Zaved., Radioelektronika 13*, 1291-1296 (1970).
25. Balakhanov, V. Ya., V. K. Zhivotov, M. F. Krotov, V. D. Rusanov, A. R. Striganov, and A. V. Titov, *Zh. Tekh. Fiz. 40*, 445-449 (1970) [*Sov. Phys. Tech. Phys. 15*, 340 (1970)].
26. Balakhanov, V. Ya., V. A. Vershkov, V. K. Zhivotov, O. A. Zinov'ev, V. D. Rusanov, and A. V. Titov, *Zh. Tekh. Fiz. 41*, 2233-2235 (1971) [*Sov. Phys. Tech. Phys. 16*, 1772 (1972)].
27. Balakhanov, V. Ya., V. K. Zhivotov, M. F. Krotov, A. R. Striganov, and M. D. Suleǐmanov, *Opt. Spektrosk. 36*, 572-578 (1974) [*Opt. Spectrosc. (USSR) 36*, 331 (1974)].
28. Barkov, A. P., P. Ya. Burchenko, E. D. Volkov, A. A. Kutsyn, M. E. Maznichenko, O. S. Pavlichenko, and K. S. Rubtsov, in: *Third International Symposium on Toroidal Plasma Confinement*, Garching, 1973, Report E-18.

29. Barkov, A. P., P. Ya. Burchenko, E. D. Volkov, V. A. Rudakov, and F. F. Tereshchenko, *Zh. Tekh. Fiz. 43*, 1170-1173 (1973) [*Sov. Phys. Tech. Phys. 18*, 743 (1973).

30. Barkov, A. P., P. Ya. Burchenko, E. D. Volkov, V. G. Konovalov, I. I. Konovalov, A. I. Ranyuk, K. S. Rubtsov, V. A. Rudakov, and A. S. Slavnyĭ, *Fiz. Plazmy 2*, 1013-1017 (1976) [*Sov. J. Plasma Phys. 2*, 562 (1976)].

31. Bass, F. G., Ya. B. Faĭnberg, and V. D. Shapiro, *Zh. Eksp. Teor. Fiz. 49*, 329-334 (1965) [*Sov. Phys. JETP 22*, 230 (1966)].

32. Butt, E. P., R. Carruthers, et al., in: *Physics of Hot Plasmas and Thermonuclear Reactions* [in Russian], Vol. 2, Izd. GU po Ispol'z. Atom. Energii pri SM SSSR (1959), pp. 370-406.

33. Berger, N. K., V. S. Koĭdan, A. N. Papyrin, A. G. Ponomarenko, V. N. Stibunov, and B. A. Yablochkov, *Zh. Eksp. Teor. Fiz. 56*, 1463-1479 (1969) [*Sov. Phys. JETP 29*, 784 (1969)].

34. Berezin, A. K., G. P. Berezina, L. I. Bolotin, and Ya. B. Faĭnberg, *At. Energ. 14*, 249-256 (1963).

35. Berezin, A. B., L. V. Dubovoĭ, and B. V. Lyublin, *Zh. Tekh. Fiz. 41*, 2323-2326 (1971) [*Sov. Phys. Tech. Phys. 16*, 1844 (1972)].

36. Berezin, A. B., L. V. Dubovoĭ, and B. V. Lyublin, in: *Plasma Physics and Controlled Nuclear Fusion Research, 2*, (IAEA) Unipub, New York ·(1971), pp. 99-109.

37. Blinov, P. I., and L. P. Zakatov, *At. Energ. 19*, 233-238 (1965).

38. Bobrovskiĭ, G. A., É. I. Kuznetsov, and K. A. Razumova, *Zh. Eksp. Teor. Fiz. 59*, 1103-1114 (1970) [*Sov. Phys. JETP 32*, 599 (1971)].

39. Bogdanov, E. V., V. Ya. Kislov, and Z. S. Chernov, *Radiotekh. Elektron. 5*, 228-238 (1960).

40. Borisov, M. D., V. A. Suprunenko, E. A. Sukhomlin, E. D. Volkov, report to the First Conference of the Physicotechnical Institute of the Academy of Sciences of the Ukrainian SSR on Plasma Physics and the Problem of Controlled Thermonuclear Reactions, June 25-July 2, 1959, Khar'kov; in: *Plasma Physics and the Problems of Controlled Thermonuclear Fusion* [in Russian], Izd. AN UkrSSR (1962), pp. 133-138.

41. Braginskiĭ, S. I., in: *Reviews of Plasma Physics*, Vol. 1, Consultants Bureau, New York (1965).

42. Budker, G. I., *At. Energ. 5*, 9-19 (1956).

43. Burchenko, P. Ya., B. T. Vasilenko, E. D. Volkov, R. M. Nikolaev, V. A. Potapenko, and V. T. Tolok, *Pis'ma Zh. Eksp. Teor. Fiz. 3*, 243-246 (1966) [*JETP Lett. 3*, 156 (1966)].
44. Burchenko, P. Ya., B. T. Vasilenko, E. D. Volkov, R. M. Nikolaev, V. A. Potapenko, and V. T. Tolok, in: *Interaction of Charged Particles with Plasmas* [in Russian], Izd. AN UkrSSR, Kiev (1967), pp. 132-138.
45. Burchenko, P. Ya., B. T. Vasilenko, E. D. Volkov, R. M. Nikolaev, V. A. Potapenko, V. A. Rudakov, F. F. Tereshchenko, V. T. Tolok, and O. S. Pavlichenko, Preprint No. 190, Physicotechnical Institute, Academy of Sciences of the Ukrainian SSR, Khar'kov, 1968 [in Russian].
46. Burchenko, P. Ya., B. T. Vasilenko, E. D. Volkov, O. S. Pavlichenko, V. A. Potapenko, V. A. Rudakov, F. F. Tereshchenko, V. T. Tolok, V. M. Zalkind, V. G. Zykov, V. I. Karpukhin, and N. I. Rudnev, in: *Plasma Physics and Controlled Nuclear Fusion, 1,* (IAEA) Unipub, New York (1969), pp. 543-560.
47. Burchenko, P. Ya., B. T. Vasilenko, E. D. Volkov, V. A. Potapenko, and V. T. Tolok, *At. Energ. 27*, 164-165 (1969).
48. Burchenko, P. Ya., B. T. Vasilenko, E. D. Volkov, V. A. Potapenko, and V. A. Rudakov, in: *Plasma Physics and the Problems of Controlled Thermonuclear Fusion* [in Russian], Izd. AN UkrSSR (1971), pp. 194-197.
49. Burchenko, P. Ya., E. D. Volkov, and V. A. Rudakov, *Zh. Tekh. Fiz. 41*, 428-430 (1971) [*Sov. Phys. Tech. Phys. 16*, 323 (1971)].
50. Burchenko, P. Ya., E. D. Volkov, V. L. Sizonenko, and K. N. Stepanov, in: *Plasma Physics and Controlled Nuclear Fusion Research, 3,* (IAEA) Unipub, New York (1971), 119-130.
51. Burchenko, P. Ya., E. D. Volkov, N. A. Manzyuk, V. A. Suprunenko, and E. A. Sukhomlin, *Zh. Tekh. Fiz. 42*, 1540-1541 (1972) [*Sov. Phys. Tech. Phys. 17*, 1232 (1973).
52. Vasil'ev, M. P., A. G. Dikiĭ, A. E. Kulaga, A. P. Litvinov, N. F. Perepelkin, and V. A. Suprunenko, *Vopr. Atom. Nauki i Tekhniki* No. 1, 40-43 (1974).
53. Vasil'ev, M. P., A. E. Kulaga, N. F. Perepelkin, G. L. Iosel'son, and V. I. Chepizhko, in: *Research on Thermal and Temperature Measurements (Diagnostics and High-Temperature Measurements). Vol. 202.*

Proceedings of the Metrological Institutes of the USSR [in Russian] (1976), pp. 61-66.
54. Vasil'ev., M. P., G. V. Zelenin, G. L. Iosel'son, V. D. Kotsubanov, A. E. Kulaga, A. P. Litvinov, N. F. Perepelkin, and V. I. Chepizhko, in: *Plasma Diagnostics* [in Russian], 4(1), Energoizdat, Moscow (1981), pp. 110-121.
55. Vedenov, A. A., E. P. Velikhov, and R. Z. Sagdeev, *Nucl. Fusion*, Supplement *2*, 465-475 (1962).
56. Vedenov, A. A., and L. I. Rudakov, *Dokl. Akad. Nauk SSSR 159*, 767-775 (1964) [*Sov. Phys. Dokl. 9*, 1073 (1965)].
57. Vedenov, A. A., and D. D. Ryutov, in: *Reviews of Plasma Physics*,*6*, Consultants Bureau, New York (1974).
58. Vekshtein, G. E., and R. Z. Sagdeev, *Pis'ma Zh. Eksp. Teor. Fiz. 11*, 297-300 (1970) [*JETP Lett. 11*, 194 (1970)].
59. Galaktionov, B. V., M. M. Larionov, and K. A. Podushnikova, *Zh. Tekh. Fiz. 35*, 172-174 (1965) [*Sov. Phys. Tech. Phys. 10*, 140 (1965)].
60. Galeev, A. A., and R. Z. Sagdeev, in: *Reviews of Plasma Physics, 7*, Consultants Bureau, New York.
61. Galeev, A. A., and R. Z. Sagdeev, in: *Reviews of Plasma Physics*, *7*, Consultants Bureau, New York.
62. Ginzburg, V. L., and V. V. Zheleznyakov, *Astron. Zh. 35*, 694-712 (1958).
63. Ginzburg, V. L., and A. A. Rukhadze, *Waves in Magnetized Plasmas* [in Russian], Nauka, Moscow (1970).
64. Golant, V. E., and A. D. Piliya, *Usp. Fiz. Nauk 104*, 413-455 (1971) [*Sov. Phys. Usp. 14*, 413 (1972)].
65. Gorbunov, L. M., V. V. Pustovalov, and V. P. Silin, *Zh. Eksp. Teor. Fiz. 47*, 1437-1453 (1964) [*Sov. Phys. JETP 20*, 967 (1965)].
66. Gordeev, G. V., *Zh. Eksp. Teor. Fiz. 27*, 19-23 (1954).
67. Gorokhov, N. A., and G. G. Dolgov-Savel'ev, *Prib. Tekh. Eksp.* No. 1, 126 (1966).
68. Gorokhov, N. A., *At. Energ. 24*, 168-172 (1968).
69. Gribkov, V. A., M. I. Ivanov, G. I. Kalinnikov, O. N. Krokhin, O. V. Kozlov, V. Ya. Nikulin, and G. V. Sklizkov, *Fiz. Plazmy 3*, 175-177 (1977) [*Sov. J. Plasma Phys. 3*, 102 (1977)].
70. Grigor'eva, L. I., A. V. Longinov, A. I. Pyatak, V. L. Sizonenko, B. I. Smerdov, K. N. Stepanov, and V. V. Chechkin, in: *Plasma Physics and Controlled Nuclear Fusion Research*, *3*, (IAEA), Unipub,

New York (1971), pp. 573-595.

71. Degtyarev, L. M., and V. E. Zakharov, *Pis'ma Zh. Eksp. Teor. Fiz.* *20*, 365-371 (1974) [*JETP Lett.* *20*, 164 (1974)].

72. Demidov, B. A., N. I. Elagin, D. D. Ryutov, and S. D. Fanchenko, *Zh. Eksp. Teor. Fiz.* *48*, 454-463 (1965) [*Sov. Phys. JETP.* *2*, 302 (1965)].

73. Demidov, B. A., and S. D. Fanchenko, *Pis'ma Zh. Eksp. Teor. Fiz.* *2*, 533-537 (1965) [*JETP Lett.* *2*, 332 (1965)].

74. Demidov, B. A., and S. D. Fanchenko, *At. Energ.* *20*, 516-518 (1966).

75. Demidov, B. A., N. I. Elagin, and S. D. Fanchenko, *Dokl. Akad. Nauk SSSR* *174*, 327-329 (1967).

76. Demidov, B. A., and S. D. Fanchenko, in: *International Conference on Plasma Confinement in Closed Systems*, Dubna, 1969 [in Russian], pp. 20-22.

77. Demirkhanov, R. A., A. G. Gevorkov, A. P. Popov, and G. I. Zverev, *Zh. Tekh. Fiz.* *30*, 306-314 (1960) [*Sov. Phys. Tech. Phys.* *5*, 282 (1960)].

78. Dikiĭ, A. G., V. M. Zalkind, G. V. Zelenin, V. G. Konovalov, V. D. Kotsubanov, A. P. Litvinov, O. S. Pavlichenko, N. F. Perepelkin, N. P. Ponomarenko, V. A. Suprunenko, V. T. Tolok, and V. M. Tonkopryad, *Pis'ma Zh. Eksp. Teor. Fiz.* *14*, 366 (1971) [*JETP Lett.* *14*, 245 (1971)].

79. Dikiĭ, A. G., O. S. Pavlichenko, N. F. Perepelkin, and V. A. Suprunenko, *Vopr. Atom. Nauki i Tekhniki*, No. 1, 22-27 (1974).

80. Dikiĭ, A. G., Yu. K. Kuznetsov, O. S. Pavlichenko, V. K. Pashnev, N. F. Perepelkin, V. A. Suprunenko, and V. T. Tolok, in: *Plasma Physics and Controlled Nuclear Fusion Research*, *2*, (IAEA), Unipub, New York (1975), pp. 45-61.

81. Dreicer, H., in: *Physics of Hot Plasmas and Thermonuclear Reactions* [Russian translation], Vol. 2, Izd. GU po Ispol'z. Atom. Energii pri SM SSSR (1959), pp. 170-180.

82. Dubovoĭ, L. V., B. A. Ivanov, and V. I. Chernobrovin, *Zh. Eksp. Teor. Fiz.* *58*, 14-25 (1970) [*Sov. Phys. JETP* *31*, 8 (1970)].

83. Dubovoĭ, L. V., A. V. Fedyakov, and A. V. Komin, *Zh. Eksp. Teor. Fiz.* *62*, 1335-1344 (1972) [*Sov. Phys. JETP* *35*, 703 (1972)].

84. Elagin, N. I., and S. D. Fanchenko, *Zh. Eksp. Teor. Fiz. 67*, 1355-1363 (1974) [*Sov. Phys. JETP 40*, 674].
85. Zheleznyakov, V. V., *Radiotekh. Elektron. 1*, 840-848 (1956).
86. Zheleznyakov, V. V., *Radio Emission of the Sun and Planets*, [in Russian], Nauka, Moscow (1964).
87. Zhivotov, V. K., G. I. Kalashnikov, A. S. Kingsep, M. F. Krotov, V. D. Rusanov, and Yu. V. Tarasov, *Zh. Eksp. Teor. Fiz. 67*, 84-88 (1974) [*Sov. Phys. JETP 40*, 43].
88. Zhil'tsov, V. A., A. N. Karkhov, V. Kh. Likhtenshtein, and D. A. Panov, in: *Plasma Physics and Controlled Nuclear Fusion Research*, 2, (IAEA) Unipub, New York (1971), pp. 75-84.
89. Zavoiskii, E. K., and V. A. Skoryupin, *Zh. Eksp. Teor. Fiz. 40*, 426-432 (1961) [*Sov. Phys. JETP 13*, 292 (1961)].
90. Zavoiskii, E. K., S. L. Nedoseev, and L. I. Rudakov, *Pis'ma Zh. Eksp. Teor. Fiz. 6*, 951-953 (1967) [*JETP Lett. 6*, 367 (1967)].
91. Zavoiskii, E. K., and L. I. Rudakov, *At. Energ. 23*, 417-431 (1967).
92. Zavoiskii, E. K., S. P. Nedoseev, L. I. Rudakov, V. D. Rusanov, V. A. Skoryupin, and S. D. Fanchenko, in: *Plasma Physics and Controlled Nuclear Fusion Research*, 2, (IAEA) Unipub, New York (1969), pp. 679-691.
93. Zavoiskii, E. K., B. A. Demidov, Yu. G. Kalinin, A. G. Plakhov, L. I. Rudakov, V. D. Rusanov, V. A. Skoryupin, G. E. Smolkin, A. V. Titov, S. D. Fanchenko, V. V. Shapkin, and G. V. Sholin, in: *Plasma Physics and Controlled Nuclear Fusion Research*, 2, (IAEA) Unipub, New York (1971), pp. 3-24.
94. Zavoiskii, E. K., Yu. K. Kalinin, V. A. Skoryupin, V. V. Shapkin, and G. V. Sholin, *Pis'ma Zh. Eksp. Teor. Fiz. 13*, 19-23 (1971) [*JETP Lett. 13*, 12 (1971)].
95. Zavoiskii, E. K., A. S. Kingsep, V. D. Ryutov, and V. A. Skoryupin, *Zh. Eksp. Teor. Fiz. 60*, 1320-1325 (1971) [*Sov. Phys. JETP. 33*, 713 (1971)].
96. Zagorodnikov, S. P., G. E. Smolkin, E. A. Striganova, and G. V. Sholin, *Pis'ma Zh. Eksp. Teor. Fiz. 11*, 475-478 (1970) [*JETP Lett. 11*, 323 (1970)].

97. Zaĭdel', A. N., G. M. Malyshev, E. Ya. Shreĭder, A. B. Berezin, V. A. Belyaeva, V. I. Gladushchak, V. V. Skidan, and L. V. Sokolova, *Zh. Tekh. Fiz. 30*, 1422-1432 (1960) [*Sov. Phys. Tech. Phys. 5*, 1346 (1961)].
98. Zaĭdel', A. N., G. M. Malyshev, E. I. Moskalev, E. A. Ptitsyna, L. V. Sokolova, and G. I. Chashchina, *Zh. Tekh. Fiz. 30*, 1433-1436 (1960) [*Sov. Phys. Tech. Phys. 5*, 1356 (1961)].
99. Zakharov, V. E., *Zh. Eksp. Teor. Fiz. 62*, 1745-1759 (1972) [*Sov. Phys. JETP 35*, 908 (1972)].
100. Zelenin, G. V., A. A. Kutsyn, M. E. Maznichenko, O. S. Pavlichenko, and V. A. Suprunenko, *Zh. Eksp. Teor. Fiz. 58*, 1879-1883 (1970) [*Sov. Phys. JETP 31*, 1009 (1970)].
101. Zykov, V. G., N. G. Sinitsa, I. A. Stepanenko, V. T. Tolok, and K. D. Sinel'nikov, *Zh. Tekh. Fiz. 34*, 1417-1423 (1964) [*Sov. Phys. Tech. Phys. 9*, 1094 (1965)].
102. Ivanov, A. A., and D. D. Ryutov, *Zh. Eksp. Teor. Fiz. 48*, 684-690 (1965) [*Sov. Phys. JETP 21*, 451 (1965)].
103. Ivanov, A. A., N. F. Perepelkin, V. V. Starykh, and S. D. Fanchenko, *Zh. Eksp. Teor. Fiz. 62*, 1362-1368 (1972) [*Sov. Phys. JETP 35*, 717 (1972)].
104. Ivanov, M. I., V. I. Kazacha, O. V. Kozlov, V. D. Rusanov, S. S. Sobolev, and A. V. Chechkin, *Zh. Tekh. Fiz. 45*, 2240-2243 (1975) [*Sov. Phys. Tech. Phys. 20*, 1404 (1975)].
105. Ioffe, M. S., R. I. Sobolev, V. G. Tel'kovskiĭ, and E. E. Yushmanov, *Zh. Eksp. Teor. Fiz. 39*, 1602-1611 (1960) [*Sov. Phys. JETP 12*, 1117 (1961)].
106. Iskol'skiĭ, A. M., R. Kh. Kurtmulaev, Yu. E. Nesterikhin, and A. G. Ponomarenko, *Zh. Eksp. Teor. Fiz. 47*, 774-776 (1964) [*Sov. Phys. JETP 20*, 517 (1965)].
107. Kadomtsev, B. B., and V. I. Petviashvili, *Zh. Eksp. Teor. Fiz. 43*, 2234-2244 (1962) [*Sov. Phys. JETP 16*, 1578 (1963)].
108. Kadomtsev, B. B., in: *Reviews of Plasma Physics, 2*, Consultants Bureau, New York (1966).
109. Kadomtsev, B. B., in: *Reviews of Plasma Physics, 4*, Consultants Bureau, New York (1966).
110. Kadomtsev, B. B., and O. P. Pogutse, *Zh. Eksp. Teor. Fiz. 53*, 2025-2033 (1967) [*Sov. Phys. JETP 26*, 1146 (1968).
111. Kadomtsev, B. B., *Collective Phenomena in Plasmas* [in Russian], Nauka, Moscow (1976).
112. Kalinin, Yu. G., D. N. Lin, V. D. Ryutov, and V. A. Skoryupin, *Zh. Eksp. Teor. Fiz. 55*, 115-121 (1968) [*Sov. Phys. JETP 28*, 61 (1969)].

113. Kalinin, Yu. G., D. N. Lin, L. I. Rudakov, V. D. Ryutov, and V. A. Skoryupin, *Dokl. Akad. Nauk SSSR 189*, 284-287 (1969) [*Sov. Phys. Dokl. 14*, 1074 (1970)].
114. Kalinin, Yu. G., D. N. Lin, V. D. Ryutov, and V. A. Skoryupin, *Zh. Eksp. Teor. Fiz. 56*, 462-471 (1969) [*Sov. Phys. JETP 29*, 252 (1969)].
115. Kalinin, Yu. G., A. S. Kingsep, D. N. Lin, V. D. Ryutov, and V. A. Skoryupin, *Zh. Eksp. Teor. Fiz. 58*, 68-75 (1970) [*Sov. Phys. JETP 31*, 38 (1970)].
116. Kalinin, Yu. G., D. N. Lin, L. I. Rudakov., V. D. Ryutov, and V. A. Skoryupin, *Zh. Eksp. Teor. Fiz. 59*, 1056-1066 (1970) [*Sov. Phys. JETP 32*, 573 (1971)].
117. Karchevskiĭ, A. I., V. G. Averin, and V. N. Bezmel'-nitsyn, *Zh. Eksp. Teor. Fiz. 10*, 26-30 (1969) [*Sov. Phys. JETP 10*, 17 (1969)].
118. Kingsep, A. S., *Zh. Eksp. Teor. Fiz. 58*, 1040-1045 (1970) [*Sov. Phys. JETP 11*, 559 (1970)].
119. Kirillov, V. D., B. A. Trubnikov, and S. A. Trushin, *Fiz. Plazmy 1*, 218-237 (1975) [*Sov. J. Plasma Phys. 1*, 117 (1975)].
120. Kovrizhnykh, L. M., *Zh. Eksp. Teor. Fiz. 51*, 1795-1810 (1966) [*Sov. Phys. JETP 24*, 1210 (1967)].
121. Kovrizhnykh, L. M., and A. A. Rukhadze, *Zh. Eksp. Teor. Fiz. 38*, 850 (1960) [*Sov. Phys. JETP 11*, 615 (1960)].
122. Koĭdan, V. S., A. N. Papyrin, A. G. Ponomarenko, and B. A. Yablochkov, *Pis'ma Zh. Eksp. Teor. Fiz. 8*, 389-393 (1968) [*JETP Lett. 8*, 241 (1968)].
123. Colgate, S. A.,I. P. Ferguson, H. P. Furth, and R. E. Wright, in: *Physics of Hot Plasmas and Thermonuclear Reactions* [Russian translation], *2*, Izd. GU po Ispol'z. Atom. Energii pri SM SSSR, Moscow (1959), pp. 449-457.
124. Korolev, O. S., and V. I. Petviashvili, *Fiz. Plazmy 1*, 436-440 (1975) [*Sov. J. Plasma Phys. 1*, 241 (1975)].
125. Kurtmulaev, R. Kh., Yu. E. Nesterikhin, V. I. Pil'skii, and R. Z. Sagdeev, in: *Plasma Physics and Controlled Nuclear Fusion Research*, *2*, (IAEA), Unipub, New York (1966), pp. 367-387.
126. Larionov, M. M., in: *Plasma Diagnostics* [in Russian], *1*, Atomizdat, Moscow (1963), pp. 132-144.
127. Larionov, M. M., *Zh. Tekh. Fiz. 34*, 1843-1851 (1964) [*Sov. Phys. Tech. Phys. 9*, 1422 (1965)].
128. Larionov, M. M., V. V. Rozhdestvenskiĭ, and D. D. Piliya, *Zh. Tekh. Fiz. 37*, 1471-1476 (1967) [*Sov. Phys. Tech. Phys. 12*, 1067 (1968)].

129. Larionov, M. M., and V. V. Rozhdestvenskiĭ, *Zh. Tekh. Fiz. 37*, 1466-1470 (1967) [*Sov. Phys. Tech. Phys. 12*, 1063 (1968)].
130. Larionov, M. M., V. A. Rodichkin, V. V. Rozhdestvenskiĭ, and A. M. Timonin, in: *Plasma Physics and Controlled Nuclear Fusion, 1*, (IAEA), Unipub, New York (1969), pp. 225-234.
131. Lovetskiĭ, E. E., and A. A. Rukhadze, *Nucl. Fusion 6*, 9-16 (1966).
132. Longinov, A. V., N. F. Perepelkin, and V. A. Suprunenko, *Fiz. Plazmy 2*, 626-636 (1976) [*Sov. J. Plasma Phys. 2*, 344 (1976)].
133. Lutsenko, E. I., Ya. B. Faĭnberg, V. A. Vasil'chuk, and N. P. Shepelev, *Zh. Eksp. Teor. Fiz. 57*, 1575-1584 (1969) [*Sov. Phys. JETP 30*, 853 (1970)].
134. Malykh, N. I., N. F. Perepelkin, L. A. Utkina, and E. S. Yampol'skiĭ, *At. Energ. 25*, 3-9 (1968).
135. Manzyuk, N. A., V. A. Suprunenko, E. A. Sukhomlin, and A. M. Ternopol, *Zh. Eksp. Teor. Fiz. 58*, 551-557 (1970) [*Sov. Phys. JETP 31*, 296 (1970)].
136. Manzyuk, N. A., N. I. Reva, V. A. Suprunenko, E. A. Sukhomlin, A. M. Ternopol, and V. T. Tolok, in: *Plasma Physics and Controlled Nuclear Fusion Research, 2*, (IAEA), Unipub, New York (1971), pp. 85-98.
137. Masud, M. M., M. E. Maznichenko, N. I. Reva, V. A. Suprunenko, E. A. Sukhomlin, A. M. Ternopol, and V. I. Tyupa, *Ukr. Fiz. Zh. 14*, 1820-1826 (1969).
138. Mikhailovskiĭ, A. B., *Zh. Tekh. Fiz. 35*, 1945-1959 (1965) [*Sov. Phys. Tech. Phys. 10*, 1498 (1966)].
139. Nezlin, M. V., *Usp. Fiz. Nauk 102*, 105-139 (1970) [*Sov. Phys. Usp. 13*, 608 (1971)].
140. Parail, V. V., and O. P. Pogutse, Preprint IAÉ 26-09 [in Russian], Moscow, 1976; in: *Voprosy Teorii Plazmy* [in Russian], 11, Moscow (1982), pp. 5-55.
141. Perepelkin, N. F., *Pis'ma Zh. Eksp. Teor. Fiz. 3*, 258-261 (1966) [*JETP Lett. 3*, 165 (1966)].
142. Perepelkin, N. F., and S. D. Fanchenko, *Dokl. Akad. Nauk SSSR 189*, 288-291 (1969) [*Sov. Phys. Dokl. 14*, 1078 (1970)].
143. Petviashvili, V. I., *Zh. Eksp. Teor. Fiz. 44*, 1993-2000 (1963) [*Sov. Phys. JETP 17*, 1340 (1963)].
144. Rozhkov, A. M., K. N. Stepanov, V. A. Suprunenko, V. I. Farenik, and V. V. Vlasov, *Ukr. Fiz. Zh. 14*, 1856-1861 (1969).
145. Romanov, Yu. L., and G. F. Filippov, *Zh. Eksp. Teor. Fiz. 40*, 123-132 (1961) [*Sov. Phys. JETP 13*, 87

(1961)].
146. Rose, D. J., and M. Clark, *Plasmas and Controlled Fusion*, MIT Press, Cambridge, Mass. (1961).
147. Rudakov, L. I., and L. V. Korablev, *Zh. Eksp. Teor. Fiz. 50*, 220-231 (1966) [*Sov. Phys. JETP 23*, 145 (1966)].
148. Rudakov, L. I., in: *Plasma Physics and Controlled Nuclear Fusion Research, 2*, (IAEA) Unipub, New York (1971), pp. 235-245.
149. Rudakov, L. I., *Dokl. Akad. Nauk SSSR 207*, 821-823 (1972) [*Sov. Phys. Dokl. 17*, 1166 (1973)].
150. Ryutov, D. D., *Dokl. Akad. Nauk SSSR 173*, 73-75 (1967) [*Sov. Phys. Dokl. 12*, 223 (1967)].
151. Sizonenko, V. L., and K. N. Stepanov, *Zh. Eksp. Teor. Fiz. 56*, 316-332 (1969) [*Sov. Phys. JETP 29*, 174 (1969)].
152. Sizonenko, V. L., K. N. Stepanov, V. A. Suprunenko, E. A. Sukhomlin, and V. T. Tolok, in: *Third European Conference on Controlled Thermonuclear Fusion*, Utrecht, The Netherlands, 1969, Report 108 [in Russian].
153. Simonov, V. A., V. V. Abozovik, and V. V. Ignat'ev, in: *Third International Conference on Plasma Physics*, Culham, UK, 1965, Report 21/167 [in Russian].
154. Sinel'nikov, K. D., and B. N. Rutkevich, *Lectures on Plasma Physics* [in Russian], Izd. Khar'k. Univ., Khar'kov (1964).
155. Spitzer, L., *Physics of Fully Ionized Gases*, Interscience, New York (1956).
156. Stepanov, K. N., and A. B. Kitsenko, *Zh. Tekh. Fiz. 31*, 167-175 (1961) [*Sov. Phys. Tech. Phys. 6*, 120 (1961)].
157. Stefanovskiĭ, A. M., *Nucl. Fusion 5*, 215-227 (1965).
158. Suprunenko, V. A., E. A. Sukhomlin, E. D. Volkov, and N. I. Rudnev, *Zh. Tekh. Fiz. 31*, 1057-1063 (1961) [*Sov. Phys. Tech. Phys. 6*, 771 (1962)].
159. Suprunenko, V. A., Ya. B. Faĭnberg, V. T. Tolok, E. A. Sukhomlin, N. I. Reva, P. Ya. Burchenko, N. I. Rudnev, and E. D. Volkov, *At. Energ. 14*, 349-352 (1963).
160. Suprunenko, V. A., E. A. Sukhomlin, and N. I. Reva, *At. Energ. 17*, 83-88 (1964).
161. Suprunenko, V. A., E. A. Sukhomlin, E. D. Volkov, and N. F. Perepelkin, Preprint KhFTI 76-34, Khar'kov Physicotechnical Institute, Khar'kov, 1976 [in Russian].

162. Sukhomlin, E. A., N. I. Reva, V. A. Suprunenko, and V. T. Tolok, *Pis'ma Zh. Eksp. Teor. Fiz. 1*, 45-49 (1965) [*JETP Lett. 1*, 68 (1965)].
163. Sukhomlin, E. A., L. I. Krupnik, N. I. Reva, P. A. Demchenko, V. I. Tyupa, A. A. Sakharov, and V. A. Suprunenko, *Ukr. Fiz. Zh. 12*, 504-506 (1967).
164. Tuck, J. L., in: *Physics of Hot Plasmas and Thermonuclear Reactions* [Russian translation], *2*, Izd. GU po Ispol'z. Atom. Energii pri SM SSSR, Moscow (1959), pp. 329-369.
165. Faǐnberg, Ya. B., *At. Energ. 11*, 313-335 (1961).
166. Faǐnberg, Ya. B., and V. D. Shapiro, in: *Sixth European Conference on Thermonuclear Fusion and Plasma Physics* [in Russian], *2*, (1973), pp. 159-183.
167. Fanchenko, S. D., B. A. Demidov, N. I. Elagin, and D. D. Ryutov, *Zh. Eksp. Teor. Fiz. 46*, 497-500 (1964) [*Sov. Phys. JETP. 19*, 337 (1964)].
168. Fanchenko, S. D., B. A. Demidov, N. I. Elagin, and D. D. Ryutov, in: *Eighth International Conference on Phenomena in Ionized Gases* [in Russian], Vienna (1967), Report 4.2.39.
169. Fanchenko, S. D., B. A. Demidov, N. I. Elagin, and N. F. Perepelkin, *Dokl. Akad. Nauk SSSR 183*, 77-79 (1968) [*Sov. Phys. Dokl. 13*, 1128 (1969)].
170. Filippov, N. V., and T. I. Filippova, in: *Plasma Physics and Controlled Nuclear Fusion Research*, *2*, (IAEA), Unipub, New York (1966), pp. 405-416.
171. Khakimov, F. Kh., and V. N. Tsytovich, *Zh. Eksp. Teor. Fiz. 64*, 1261-1271 (1973) [*Sov. Phys. JETP 37*, 641 (1973)].
172. Harding, G. N., A. N. Dellis, A. Gibson, Jones, D. J. Lees, R. W. P. McWhirter, S. A. Ramsden, and S. Ward, in: *Physics of Hot Plasmas and Thermonuclear Reactions* [Russian translation], *2*, Izd. GU po Ispol'z. Atom. Energii pri SM SSSR, Moscow (1959), pp. 652-674.
173. Kharchenko, I. F., Ya. B. Faǐnberg, R. M. Nikolaev, E. A. Kornilov, E. I. Lutsenko, and N. S. Pedenko, *Zh. Eksp. Teor. Fiz. 38*, 685-692 (1960) [*Sov. Phys. JETP 11*, 493 (1960)].
174. Kharchenko, I. F., Ya. B. Faǐnberg, R. M. Nikolaev, E. A. Kornilov, E. I. Lutsenko, and A. S. Pedenko, *Zh. Tekh. Fiz. 31*, 761-769 (1961) [*Sov. Phys. Tech. Phys. 6*, 551 (1962)].
175. V. N. Tsytovich, *Nonlinear Effects in Plasma*,

Plenum, New York (1970).

176. Tsytovich, V. N., *Theory of Turbulent Plasma*, Plenum, New York (1974).

177. Shapiro, V. D., and V. I. Shevchenko, *Zh. Eksp. Teor. Fiz. 42*, 1515-1528 (1962) [*Sov. Phys. JETP 15*, 1053 (1962)].

178. Shapiro, V. D., *Zh. Eksp. Teor. Fiz. 44*, 613-625 (1963) [*Sov. Phys. JETP 17*, 416 (1963)].

179. Shapiro, V. D., *Zh. Tekh. Fiz. 31*, 522-528 (1961) [*Sov. Phys. Tech. Phys. 6*, 376 (1961)].

180. Shapiro, V. D., V. I. Shevchenko, A. A. Galeev, R. Z. Sagdeev, and Yu. F. Sigov, *Fiz. Plazmy 1*, 10-20 (1975) [*Sov. J. Plasma Phys. 1*, 5 (1975)].

181. Shklovskiĭ, I. S., *Cosmic Radio Emission* [in Russian], Gostekhizdat, Moscow (1965).

182. Sholin, G. V., *Dokl. Akad. Nauk SSSR 195*, 589-592 (1970) [*Sov. Phys. Dokl. 15*, 1040 (1971)].

183. Aamodt, R. E., and W. F. Drummond, *Plasma Phys. 6*, 147-151 (1964).

184. Adati, K., R. Sugihara, T. Uyama, and K. Watanabe, in: *Sixth European Conference on Controlled Fusion and Plasma Physics*, Moscow, 1973, *1*, pp. 561-564.

185. Adati, K., H. Iguchi, Y. Ito, T. Kawabe, K. Kondo, O. Mitarai, K. Muraoka, and R. Sugihara, *Phys. Rev. Lett. 35*, 280-284 (1975).

186. Adati, K., K. Iguchi, Y. Ito, T. Kawabe, K. Kawasaki, T. Oda, R. Sugihara, and T. Yokota, in: *Seventh European Conference on Controlled Fusion and Plasma Physics*, Lausanne, 1975, *1*, p. 166.

187. Alexeff, I., L. A. Berry, J. M. Dudley, K. Y. Estabrook, A. Hirose, W. D. Jones, R. V. Neidigh, J. P. Olsen, F. R. Scott, W. L. Stieling, M. M. Widner, and W. R. Wing, in: *Plasma Physics and Controlled Fusion Research, 2*, (IAEA), Unipub, New York (1971), pp. 221-233.

188. Amagishi, J., H. Iguchi, Y. Ito, T. Kawabe, A. Mase, and T. Tsukishima, *Phys. Lett. A63*, 31-34 (1977).

189. Amagishi, Y., A. Hirose, H. W. Piekaar, and H. M. Skaragard, in: *Plasma Physics and Controlled Fusion Research, 3*, (IAEA), Unipub, New York (1977), pp. 11-17.

190. Arunasalam, V., J. C. Hosea, and R. Cano, in: *Eighth European Conference on Controlled Fusion and Plasma Physics*, Prague, 1977, *1*, p. 137.

191. Bengtson, R. D., K. W. Gentle, J. Jancarik, S. S.

Medley, P. Nielsen, and P. Phillips, *Phys. Fluids 18*, 710-716 (1975).
192. Bernstein, W., F. F. Chen, M. A. Heald, and A. Z. Kranz, *Phys. Fluids 1*, 430-437 (1958).
193. Bernstein, I. B., E. A. Frieman, R. M. Kulsrud, and M. N. Rosenbluth, *Phys. Fluids 3*, 136-137 (1960).
194. Biskamp, D., and R. Chodura, in: *Plasma Physics and Controlled Nuclear Fusion Research, 2*, (IAEA), Unipub, New York (1971), pp. 265-276.
195. Biskamp, D., and R. Chodura, in: *Fifth European Conference on Controlled Fusion and Plasma Physics, 2*, Grenoble (1972), pp. 93-104.
196. Biskamp, D., R. Chodura, and C. T. Dum, in: *Sixth European Conference on Controlled Fusion and Plasma Physics, 1*, Moscow (1973), pp. 461-464.
197. Bocharov, V. K., V. L. Berechny, E. D. Volkov, B. V. Kravchin, V. G. Konovalow, V. I. Kononenko, V. A. Rudakow, E. A. Sukhomlin, V. A. Suprunenko, and A. M. Ternopol, in: *Eighth European Conference on Controlled Fusion and Plasma Physics, 1*, Prague (1977), p. 125.
198. Bohm, D., and E. P. Gross, *Phys. Rev. 75*, 1851-1864 (1949).
199. Bowles, K. L., *Phys. Rev. Lett. 1*, 454-455 (1958).
200. Brower, D. F., D. E. Dunaway, J. H. Malberg, C. L. Oxlay, M. Stearns, D. W. Kerst, F. R. Scott, S. P. Cunningham, and R. G. Tuckfield, *Proceedings of the Second UN International Conference on the Peaceful Uses of Atomic Energy*, Geneva, 1958, *32*, pp. 110-125.
201. Buneman, O., *Phys. Rev. 115*, 503-517 (1959).
202. Camac, M., A. R. Kantrowitz, M. M. Litvak, R. M. Patrick, and H. E. Petschek, *Nucl. Fusion, Supplement, Part 2*, 423-445 (1962).
203. Coor, T., S. P. Cunningham, R. A. Ellis, M. A. Heald, and A. Z. Kranz, *Phys. Fluids 1*, 411-420 (1958).
204. Coppi, B., F. Pegararo, R. Pozzoli, and G. Rewoldt, *Nucl. Fusion 16*, 309-328 (1976).
205. Costley, A. E., R. J. Hastie, J. W. M. Paul, and J. Chamberlain, *Phys. Rev. Lett. 33*, 758-761 (1974).
206. Demidov, B. A., N. I. Elagin, N. F. Perepelkin, and S. D. Fanchenko, Preprint IAE-1651, Kurchatov Institute of Atomic Energy, Moscow, 1968.
207. Dougherty, J. P., and D. T. Farlay, *Proc. R. Soc. Math. Phys. Sci. 259*, 79-99 (1960).
208. Dreicer, H., *Phys. Rev. 115*, 238-249 (1959).

209. Drummond, W. E., and D. Pines, *Nucl. Fusion, Supplement, Part 3*, pp. 1049-1057 (1962).
210. Drummond, W. E., and M. N. Rosenbluth, *Phys. Fluids 5*, 1507-1513 (1962).
211. Drummond, W. E., J. R. Thompson, M. L. Slaan, and H. V. Wong, in: *Plasma Physics and Controlled Nuclear Fusion Research, 2*, (IAEA), Unipub, New York (1971), pp. 167-193.
212. Ellis, R. A., L. P. Goldberg, and J. G. Gorman, *Phys. Fluids 3*, 468-473 (1960).
213. Fainberg, Ya. B., *Part. Accel. 6*, 95-106 (1975).
214. Fanchenko, S. D., B. A. Demidov, N. I. Elagin, and N. F. Perepelkin, *Phys. Rev. Lett. 21*, 789-791 (1968).
215. Fidone, J., G. Ramponi, and P. Brossier, Preprint EUR-CEA-FC-882, Fontenay-aux-Roses, France, 1977.
216. Field, E. C., and B. D. Fried, *Phys. Fluids 7*, 1937-1951 (1964).
217. Furth, H. P., J. Killeen, and M. N. Rosenbluth, *Phys. Fluids 6*, 459-484 (1963).
218. Gorbunov, E. P., D. P. Ivanov, N. J. Peacock, D. C. Robinson, and V. S. Strelkov, in: *European Conference on Controlled Fusion and Plasma Physics*, Rome, 1970, *1*, p. 19.
219. Gordon, W. E., *Proc. IREE 46*, 1824-1829 (1958).
220. Griem, H. R., *Spectral Line Broadening by Plasmas*, Wiley, New York (1974).
221. Hamberger, S. M., A. Malein, J. H. Adlam, and M. Friedman, *Phys. Rev. Lett. 19*, 350-352 (1967).
222. Hamberger, S. M., and M. Friedman, *Phys. Rev. Lett. 21*, 674-677 (1968).
223. Hamberger, S. M., and M. Friedman, Preprint CLM-P177, Culham Lab., Abingdon, 1968.
224. Hamberger, S. M., J. Jančarik, L. E. Sharp, D. A. Aldcroft, and Wetherell, in: *Plasma Physics and Controlled Fusion Research, 2* (IAEA), Unipub, New York (1971), pp. 37-53.
225. Hamberger, S. M., and J. Jančarik, *Phys. Fluids 15*, 825-836 (1972).
226. Hamberger, S. M., L. E. Sharp, and J. Jančarik, in: *Fifth European Conference on Controlled Fusion and Plasma Physics*, Grenoble, 1972, *1*, p. 146.
227. Hamberger, S. M., L. E. Sharp, A. Molein, J. Bradley, and W. McKay, in: *Plasma Physics and Controlled Fusion Research, 2* (IAEA), Unipub, New York (1975), pp. 21-32.
228. Harrison, E. R., *Phil. Mag. 3*, 1318-1325 (1958).

229. Harrison, E. R., *Plasma Phys. 4*, 7-13 (1962).
230. Heald, M. D., USAEC Report TJD-7520, Vol. 1, 1956, pp. 202-209.
231. Hirose, A., I. Alexeff, W. D. Jones, S. T. Kush, and K. E. Lonngren, *Phys. Rev. Lett. 25*, 1563-1567 (1970).
232. Hirose, A., S. Q. Mah, H. M. Skarsgard, A. R. Strilchuk, and D. W. A. Whitfield, *Phys. Rev. Lett. 28*, 1185-1188 (1972).
233. Hirose, A. T. Kawabe, and H. M. Skarsgard, in: *Proceedings of the Third International Symposium on Toriodal Plasma Confinement*, Garching, 1973, Paper E-2.
234. Hirose, A., Y. Amagishi, M. R. Shubaly, H. M. Skarsgard, and B. F. White, in: *Plasma Physics and Controlled Fusion Research, 3* (IAEA), Unipub, New York (1975), pp. 299-306.
235. Hirose, A., H. W. Piekaar, and H. M. Skarsgard, *Nucl. Fusion 16*, 963-969 (1976).
236. Ichimaru, S., and S. H. Starr, *Phys. Rev. A2*, 821-825 (1970).
237. Ikezi, H., R. P. H. Chang, and R. A. Stern, *Phys. Rev. Lett. 36*, 1047-1051 (1976).
238. Ishii, K., and T. Kawabe, *Ann. Rev. Nagoya Univ.*, 50-51 (April 1974-March 1975).
239. Jensen, T. H., and F. R. Scott, *Phys. Fluids 11*, 1809-1819 (1968).
240. Johnson, J., J. Greene, and B. Coppi, *Phys. Fluids 6*, 1169-1183 (1963).
241. Kadomtsev, B. B., and O. P. Pogutse, *Nucl. Fusion 11*, 67-92 (1971).
242. Kalfsbeek, H. W., B. De Groot, T. G. A. Winkel, H. De Kluiver, and H. W. Piekaar, in: *Seventh European Conference on Controlled Fusion and Plasma Physics*, Lausanne, 1975, *1*, p. 163.
243. Kawamura, T., and T. Kawabe, in: *Proceedings of the Third International Meeting on the Theoretical and Experimental Aspects of Toroidal Plasma Heating*, Grenoble, 1976, *2*, pp. 179-203.
244. Kim, H. C., R. L. Stenzel, and A. Y. Wong, *Phys. Rev. Lett. 33*, 886-889 (1974).
245. De Kluiver, H. W., Report 77-106, Assoc. Euratom-Fom. FOM-Inst. voor Plasma-fysice, Jutphaas, 1977.
246. Langmuir, I., *Phys. Rev. 26*, 585-613 (1925).
247. Lashmore-Davies, C. N., *Nucl. Fusion 15*, 213-216 (1975).
248. Lavrovsky, V. A., V. M. Deyev, S. A. Rogashkov, Yu.

B. Yaremenko, and I. F. Kharchenko, in: *Eighth Conference on Ionized Gases*, Vienna, 1967, *1*, p. 375.
249. Looney, D. H., and S. C. Brown, *Phys. Rev. 93*, 965-969 (1954).
250. Luppi, R., and M. Martone, in: *Plasma Physics and Controlled Fusion, 1*, (IAEA), Unipub, New York (1975), pp. 227-230.
251. Mah, S. Q., H. M. Skarsgard, and A. R. Strilchuk, *Phys. Rev. Lett. 25*, 1409-1413 (1970).
252. Martin, D. F., *Proc. R. Soc. 193*, 44-59 (1948).
253. Martone, M., in: *Sixth European Conference on Controlled Fusion and Plasma Physics*, Moscow, 1974, *2*, pp. 324-325.
254. Mase, A., T. Tsukishima, H. Iguchi, Y. Ito, and T. Kawabe, Research Report, Institute of Plasma Physics, Nagoya Univ., IPPJ-291, 1977.
255. Merrill, H. J., and H. W. Webb, *Phys. Rev. 55*, 1191-1198 (1939).
256. Mills, R. G., MATT-150, Plasma Phys. Lab Rept., Princeton Univ., 1974.
257. Nakagawa, Y., M. Tanikawa, K. Watanabe, K. Adati, H. Iguchi, K. Ishii, Y. Ito, J. Jacquinot, T. Kawabe, T. Kawamura, K. Muraoka, T. Oda, and R. Sugihara, in: *Plasma Physics and Controlled Fusion Research, 3*, IAEA, Vienna, 1975, pp. 287-297.
258. Perkin, R. M., A. D. Craig, L. S. Holms, and J. W. M. Paul, in: *Plasma Physics and Controlled Nuclear Fusion Research, 2*, IAEA, 1975, p. 681.
259. Paul, J. W. M., L. S. Holms, M. J. Parkinson, J. Sheffield, *Bull. Am. Phys. Soc. 10*, 223-233 (1965).
260. Payne-Scott, R., D. E. Jabsley, and J. G. Bolton, *Nature 160*, No. 4060, pp. 256-257 (1947).
261. Penrose, O., *Phys. Fluids 3*, 258-265 (1960).
262. Perepelkin, N. F., V. A. Suprunenko, M. P. Vasil'ev, and A. G. Diky, in: *Fifth European Conference on Controlled Fusion and Plasma Physics, 1*, Grenoble, 1972, p. 173.
263. Perepelkin, N. F., V. A. Suprunenko, M. P. Vasil'ev, and A. G. Diky, in: *Third International Symposium on Toroidal Plasma Conference*, Rept. E 19, Garching, 1973.
264. Perepelkin, N. F., V. A. Supruneko, V. I. Petviashvili, and M. P. Vasil'ev, in: *Eighth European Conference on Controlled Fusion and Plasma Physics, 1*, Prague, 1977, p. 144.

265. Piekaar, H. W., H. De Kluiver, H. Schrijver, W. R. Rutgers, P. A. A. Ambags, and B. De Groot, in: *Fifth European Conference on Controlled Fusion and Plasma Physics, 1*, Grenoble, 1972, p. 147.
266. Piekaar, H. W., *Plasma Phys. 15*, 565-575 (1973).
267. Putley, E. H., *Appl. Opt. 4*, 649-656 (1965).
268. Robinson, D. C., and M. G. Rusbridge, *Plasma Phys. 11*, 73-100 (1969).
269. Rudakov, L. I., and V. N. Tsytovitch, *Plasma Phys. 13*, 213-228 (1971).
270. Rudakov, L. I., A. S. Kingsep, and S. N. Sudan, *Phys. Rev. Lett. 31*, 1482-1484 (1973).
271. Rutgers, W. R., *Phys. Lett. A40*, 99-100 (1972).
272. Rutgers, W. R., and H. De Kluiver, *Phys. Lett. A53*, 423-425 (1975).
273. Rutgers, W. R., and H. W. Kalfsbeek, in: *Seventh European Conference on Controlled Fusion and Plasma Physics, 1*, Lausanne, 1975, p. 167.
274. Rutgers, W. R., and H. De Kluiver, *Phys. Lett. A58*, 99-101 (1976).
275. Salpeter, E. E., *Phys. Rev. 120*, 1528-1535 (1960).
276. Oomens, A. A. M., L. Th. M. Ornstein, R. R. Parker, F. C. Schüller, and R. J. Taylor, *Phys. Rev. Lett. 36*, 255-257 (1976).
277. Schrijver, H., W. R. Rutgers, I. A. Van Elst, H. A. Post, H. W. Kalfsbeek, H. W. Piekaar, and H. De Kluiver, in: *Plasma Physics and Controlled Nuclear Fusion Research, 3*, IAEA, Vienna, 1975, pp. 271-285.
278. Sharp, L. E., and S. Mrowka, in: *Seventh European Conference on Controlled Fusion and Plasma Physics, 1*, Lausanne, 1975, p. 168.
279. Shepherd, L. T., and H. M. Skarsgard, *Phys. Rev. Lett. 10*, 121-123 (1963).
280. Sizonenko, V. L., and K. N. Stepanov, *Plasma Phys. 13*, 1033-1050 (1971).
281. Skarsgard, H. M., *Can. J. Phys. 41*, 450-468 (1963).
282. Smullin, L. D., and W. D. Getty, *Phys. Rev. Lett. 9*, 3-9 (1962).
283. Spitzer, L., and R. Harm, *Phys. Rev. 89*, 977-981 (1953).
284. Stenzel, R. L., *Phys. Fluids 19*, 865-871 (1976).
285. Stephanakis, S. J., L. S. Levine, D. Mosher, I. M. Vitkovitsky, and F. Young, *Phys. Rev. Lett. 29*, 568-569 (1972).
286. Stringer, T. E., *Plasma Phys. 6*, 267-279 (1964).

287. Sturrock, P. A., *Phys. Rev. 141*, 186-191 (1966).
288. Suprunenko, V. A., E. A. Sukhomlin, and V. T. Tolok, *Plasma Phys. 12*, 627-637 (1970).
289. Suprunenko, V. A., E. A. Sukhomlin, and V. T. Tolok, *Plasma Phys. 15*, 353-367 (1973).
290. Suprunenko, V. A., E. A. Sukhomlin, E. D. Volkov, and N. F. Perepelkin, in: *Third International Meeting on Theoretical and Experimental Aspects of Heating Toroidal Plasmas*, Grenoble, 1976, *2*, pp. 407-410.
291. Tanikawa, T., T. Uyama, and K. Watanabe, *Ann. Rev. Nagoya Univ.*,(April 1971-March 1972), pp. 64-66.
292. TFR Group (presented by R. Cano), in: *Seventh European Conference on Controlled Fusion and Plasma Physics*, Lausanne, 1975, *1*, p. 14b.
293. TFR, NRL Groups (presented by A. E. Costley), in: *Seventh European Conference on Controlled Fusion and Plasma Physics*, Lausanne, 1975, *1*, p. 14a.
294. Tsytovich, V. N., in: *Annual Review of Astronomy and Astrophysics*, *11*, Annual Reviews, Palo Alto, Ca. (1973), pp. 363-386.
295. Uchida, T., K. Mujamoto, J. Fujita, C. Leloop, S. Nawasaki, N. Inoul, Y. Suzuki, and K. Adati, *Nucl. Fusion 19*, 259-265 (1969).
296. Vlasenkov, V. S., V. M. Leonov, V. G. Merezkin, and V. S. Mukhovatov, *Nucl. Fusion 13*, 509-516 (1973).
297. Volkov, E. D., E. M. Latsko, K. S. Rubtsov, and V. A. Rudakov, in: *Seventh European Conference on Controlled Fusion and Plasma Physics*, *1*, Lausanne, 1975, p. 117.
298. Watanabe, K., and K. Nishikawa, Research Report, IPPJ-193, Nagoya Univ., 1974.
299. Weinstock, J., and B. Bezzerides, *Phys. Fluids 16*, 2287-2303 (1973).
300. Wesson, J. A., *Nucl. Fusion 18*, 87-96 (1978).
301. Wharton, C. B., and J. H. Malmberg, *Phys. Fluids 11*, 2655-2664 (1968).
302. Wharton, C. B., P. Korn, and S. Robertson, *Phys. Rev. Lett. 27*, 499-501 (1971).
303. Wild, J. P., and S. F. Smerd, *Ann. Rev. Astrophys. 10*, 159-196.
304. Wort, D. J. H., *Plasma Phys. 6*, 237-250 (1964).